Statistical Auditing

By Donald M. Roberts
Associate Professor
of Business Administration
University of Illinois
at Urbana-Champaign

American Institute of Certified Public Accountants

1211 Avenue of the Americas
New York, New York 10036

1211 Avenue of the Americas, New York, N.Y. 10036
Printed in the United States of America

Contents

Foreword

This book is for auditors who want to use statistical sampling. Even a scanning of it should convince many auditors that statistical sampling is as specialized an audit tool as the computer and also that it holds nearly as much promise for enhancing audit practice. Both statistical sampling and the computer require improved audit planning and a more thorough understanding of the objectives of audit tests. However, until now, there has not been an understandable and comprehensive explanation of statistical sampling in auditing comparable to material available on the computer.

The book covers the array of statistical techniques available to the auditor thoroughly, yet understandably, and explains what they are, the assumptions on which they are based, and in what circumstances they should be used. Some statistical techniques—such as dollar unit sampling, discovery sampling, or difference estimation—have been touted as ideally suited to the needs of auditors. This book makes clear that no sampling plan is superior in every situation. More important, it explains the considerations that should determine the auditor's choice of a particular sampling plan.

The book provides practical ideas on incorporating statistical sampling into audit practice. Advice is given on establishing firm-wide policies, conducting training programs, and documenting and reviewing statistical applications. Often, valid application of statistical sampling in auditing requires specialized knowledge that exceeds the practical knowledge and skills of a staff auditor; this book will increase recognition of that. However, it will also provide the basis for a CPA firm to develop and implement the specialization required for efficient and effective use of statistical sampling in its practice.

Manual application of statistical sampling is often expensive, tedious, and time-consuming—disadvantages that usually offset the expected advantages of statistical sampling. This book not only recognizes the importance of the computer to effective and efficient application of statistical sampling, but supplies useful computer programs as well.

Finally, although some progress has been made in relating statistical tests to audit judgments, this book is more comprehensive in this area. Statistical tests often provide only part of the audit evidence for a particular account or class of transactions and never provide more than a portion of the evidence supporting an opinion on financial statements. This book offers many suggestions for integrating statistical tests with other audit tests and relating statistical judgments to other audit judgments.

Auditors have long needed a practical and knowledgeable explanation of statistical applications in auditing. Most discussions about statistics in auditing are either simplified introductions without enough detail to permit direct application or esoteric dissertations filled with statistical jargon and complex formulas. This book is neither simplistic nor esoteric; it provides a systematic approach to audit planning that includes statistics as an important tool.

Douglas R. Carmichael
Vice President—Technical Services
American Institute of Certified Public Accountants

Preface

The purpose of this book is to help the auditor use statistical sampling in audit practice. Statistical techniques currently being used in practice are discussed together with some suggestions concerning how and when each may be implemented. The major topic addressed is the integration of statistical sampling into the auditing process.

Although a brief review is provided, the reader is expected to be familiar with the basic concepts of statistical sampling. The exposition is directed to auditors who already know something about commonly used statistical techniques and want to incorporate them into an audit practice.

The necessary background in statistical concepts may have been acquired in a variety of ways: the programmed instruction series published by the AICPA, college coursework, or programs available through organizations such as the Institute of Internal Auditors, the state societies of CPAs, or one of the CPA firms. However, those persons whose backgrounds consist of self-instruction would benefit from discussion with others who have had some experience in using statistical techniques in auditing.

The first chapter discusses the audit process from the point of view of the auditor who wishes to limit audit risk and, thus, demonstrates the *role* of statistical sampling. Chapters 2 through 6 summarize the basic statistical concepts and techniques that are currently used in statistical auditing. Chapter 7 suggests procedures that might be used in integrating these techniques into the auditing process in order to limit the risk caused by observing only a sample. Chapter 8 illustrates those procedures by means of an extended case study, and chapter 9 describes the set of computer programs that assist the auditor in planning, selecting, and evaluating statistical samples. Finally, the problems of training and implementation are discussed in chapter 10.

Preparation of this book began after publication of the six-volume series, *An Auditor's Approach to Statistical Sampling.* The first five volumes of that series are programmed texts treating the basic statistical techniques; volume 6 is a field manual that illustrates, by means of case studies, tables, and time sharing computer programs, how these techniques can be applied. While these books serve the useful purpose of introducing basic statistical concepts, it was felt that something more would be needed if a practice unit were to decide that statistical sampling should be used in audit engagements.

I began work on this book in June 1974 at the AICPA in New York where I spent a delightful year as a research associate on the staff of the auditing standards division. I received much help and encouragement from my colleagues, especially Douglas R. Carmichael. Throughout the writing process, members of the Statistical Sampling Subcommittee have been active collaborators. Their suggestions and criticisms have helped to improve the book, and I wish to express my deep gratitude for their efforts.

Nevertheless, I accept the responsibility for the final product. Many subcommittee members have strongly suggested that the present level of exposition is not suitable for the majority of practitioners. I acknowledge that charge and offer in defense the plea that the subject matter is complex and attempts to make it appear simple would be dangerous. The danger is that simple solutions to complex problems are often inappropriate.

It is hoped that the reader will find some suggestions that can be put into immediate practice and others that will require thought and considerable modification before they can be implemented. The issues discussed are important for the practitioner who wants to improve usage of statistical techniques.

A special note of thanks is due to the members of the subcommittee—past and present—who gave so much of their time and effort to this project. I want to acknowledge especially the efforts of Robert B. Ilderton, who wrote all the time sharing computer programs in chapter 9; Carmen Spinelli, who is responsible for both the CPA-1 and CPA-2 programs; COMSHARE, INC., which donated computer time during the development of the program; James Kusko, who read all the early drafts and made many valuable suggestions; James K. Loebbecke, who was chairman of the subcommittee during most of the process and who personally contributed much to the book; and Robert K. Elliott, the present chairman, who thoroughly reviewed a final draft and offered many excellent suggestions.

Urbana, Illinois
October 1977

D.R.

Statistical Auditing

1

Statistical Sampling in the Audit Process

Statistics has been defined as "a body of methods for making wise decisions in the face of uncertainty."[1] Similarly, statistical auditing could be defined as a body of methods for making wise auditing decisions in the face of uncertainty. At first glance such a definition appears pretentious: After all, auditors have made wise decisions for years without the aid of statistical sampling. What, then, does statistics offer the auditor?

Roughly speaking, statistical sampling helps answer one of the auditor's three key questions concerning the nature, extent, and timing of his audit procedures. The auditor can determine the *extent* of testing more objectively when using statistical sampling in tests of details rather than judgmental samples. That is not to say that statistical sampling replaces the auditor's judgment. Rather, statistical sampling allows the auditor to exercise judgment relative to the amount of sampling risk that can be borne and to express that sampling risk quantitatively.

The problem of controlling the sampling risk that an incorrect conclusion will be reached because only a sample has been examined has been extensively studied only when a single audit procedure is considered. However, some technical statistical problems remain unresolved. These pertain to which statistical techniques may be validly used in a particular set of circumstances. Both theoretical and empirical research studies have contributed to improving the statistical techniques the auditor may use.

Auditing is a very complex process in which the auditor uses many sources of evidence—some statistical and some nonstatistical. There are strong interrelationships among the many audit procedures requiring the auditor who wishes to use them to integrate statistical tests into the general audit process.

While controlling the sampling risk for a single audit test is important, it is an even greater challenge to control the sampling risk for audit tests considered as a whole. Doing this requires careful planning of the audit program, including nonstatistical

1. See Wallis and Roberts [22], page 3.

as well as statistical procedures. It requires the auditor to think statistically—not to be a statistician, but to understand thoroughly the concepts of sampling risk and be able to apply them creatively.

The goal of this book is to help the auditor achieve that understanding. To do this the auditor needs to grapple with some difficult problems, among the most difficult of which are those that require quantification of some aspects of professional judgment. For example, what is the quantifiable likelihood that a particular set of accounting controls would prevent or detect a particular type of error?

Many auditors feel uncomfortable with the prospect of attaching numbers to these kinds of judgments; assigning a number may create a false sense of exactness.

Such an attitude is understandable. However, quantification merely makes explicit that which has always been implicit. With or without statistical sampling, the auditor has determined the extent of his tests of details, the timing of the auditing procedures, and the nature of those procedures. Consequently, an expression of some judgments on a numerical scale does not entail procedures different from those normally required—only that judgments be rendered explicitly.

Using numbers to reflect professional judgment improves an auditor's ability to communicate examination results to others. The auditor called upon to defend procedures can demonstrate their rationality and consistency. The numbers are the result of a reasoned process—the auditor's examination and evaluation. For example, while different auditors examining the same evidence may use different numerical assignments, their results would ordinarily exhibit strong similarities. Thus, if both use a numerical scale to express the maximum possible reliance, both would probably assign a low number to a weak system of internal control and a high number to a strong system.

The attitude expressed in this book is that attempting to quantify certain judgments is worthwhile as long as the inexactness of the resulting numbers is recognized. To place the role of sampling risk into perspective, the following discussion focuses on the basic audit process and the problems of developing an audit strategy. The purposes of this discussion are to demonstrate the role that sampling plays in the auditor's audit program and to highlight the relative contribution of sampling risk to overall risk.

The Basic Audit Process

The auditor uses many techniques in addition to statistical sampling to gather the evidential matter on which to base a professional opinion. The portions of the audit process that are relevant to that decision to use statistical sampling are the following:

1. System review and preliminary evaluation of internal accounting control.
2. Audit program design.
3. Application of the audit procedures, evaluation of the evidential matter, and refinement of the audit program as required.

System Review

Statistical sampling is not used to review the system of internal accounting control. However, the results of the system review and preliminary evaluation directly affect the auditor's decision to use statistical sampling in his tests of details. From the system review, the auditor obtains detailed information concerning the procedures and methods prescribed to achieve internal accounting control. The preliminary evaluation of the system of internal accounting control is made on the basis of this review and consequently is a conditional evaluation that assumes satisfactory compliance.

Section 320 of Statement on Auditing Standards (SAS) no. 1 describes the auditor's study and evaluation of internal control, and SAS no. 3 considers how electronic data processing affects it. To obtain maximum benefits from this book, the reader should be familiar with those professional pronouncements. Unless otherwise indicated, section numbers cited throughout the book refer to SAS no. 1.

Audit Program Design

The next phase of the basic audit process is to design a tentative audit program. The tentative audit program specifies, in detail, the set of audit procedures to be used to satisfy the third standard of field work concerning the sufficiency and competence of evidential matter. As section 320.70 states,

> The evidential matter required by the third standard is obtained through two general classes of auditing procedures: (a) tests of details of transactions and balances and (b) analytical review of significant ratios and trends and resulting investigation of unusual fluctuations and questionable items. These procedures are referred to . . . as "substantive tests."

In addition to specifying the substantive audit procedures, the auditor specifies the compliance tests to be conducted. As section 320.55 indicates, compliance tests are necessary when the prescribed internal control procedures are to be relied upon in determining the nature, timing, or extent of substantive tests, but are unnecessary otherwise.

Audit Risk. Uncertainty is inherent in auditing. Indeed, the general purpose of auditing procedures is to reduce the auditor's uncertainty to a tolerable level.[2] The risk the auditor faces is that material errors or irregularities, if they exist, will not be detected. The auditor is responsible for controlling this risk and exercises control by determining the nature, extent, and timing of his substantive procedures.

A major portion of the auditor's tentative decisions concerning the nature, extent, and timing of his substantive procedures depend upon his preliminary evaluation of the system of internal control. During the review of the system, the auditor considers the type of errors and irregularities that could occur and the accounting control procedures that should prevent or detect such errors and irregularities.[3] The preliminary evaluation reflects his assessment of the likelihood that each type of error or irregularity could occur in a material amount.[4] As a working hypothesis, his assessment assumes satisfactory compliance with the prescribed internal control procedures that he considers pertinent. Pertinent procedures are defined in section 320B.15 as

2. Section 330.10 states: "In the great majority of cases, the auditor finds it necessary to rely on evidence that is persuasive rather than convincing. Both the individual assertions in financial statements and the overall proposition that the financial statements as a whole present fairly, in conformity with generally accepted accounting principles, the financial position, results of operations, and changes in financial position are of such a nature that even an experienced auditor is seldom convinced beyond all doubt with respect to all aspects of the statements being examined."
3. Section 320.65.
4. Throughout this book the term *likelihood* is used to designate a probability that is subjectively determined.

"those which, if not purported to be in use, would have affected adversely the auditor's preliminary evaluation of the system prior to his tests of compliance."

The auditor's effort to reduce the risk of not detecting a material amount of error depends on an assessment of the likelihood that material errors could occur in the accounting process. The risk levels of the planned substantive tests are based on this assessment. When the auditor decides not to rely on the system, he determines the nature, extent, and timing of the substantive tests so that they alone achieve a tolerably low risk of failing to detect a material error. As reliance on the system increases, the tolerable risk that the substantive tests would fail to detect a material error is allowed to increase. This increased risk is justified whenever the planned reliance is appropriate. When the auditor relies on the system to a greater extent than he would if he knew the true effectiveness of the pertinent procedures, the risk of missing a material error is higher than is appropriate. The risk that the auditor relies on the system to a greater extent than he would if he had complete knowledge is called here the *risk of unwarranted reliance.*

Unwarranted reliance may occur when the auditor overrates the strength of the system of internal accounting control. This may happen when preliminary assessment of the likelihood that material errors could occur is too low or when tests of compliance with pertinent procedures incorrectly indicate that compliance is satisfactory. Although unwarranted reliance is not explicitly mentioned in section 320, it is implicitly recognized there. For instance, the risk that the preliminary assessment of the likelihood of material error is too low depends on both the auditor's judgment and the actual risk that material errors will occur in the accounting process.

Viewed in this way, the auditor's risk of not detecting a material error can be controlled only if he controls both the risk that substantive tests fail to detect a material amount of error and the risk of unwarranted reliance on the system of internal accounting controls.

Statistical sampling pertains only to one aspect of the total audit risk. This is the possibility that audit procedures—both compliance and substantive—restricted to a sample of details of transactions or balances might produce results that are different from those produced when the procedures are applied in the same way to all the details. This aspect, known as *sampling risk,* can be objectively measured and controlled when statistical sampling is used to determine the extent of the application of audit procedures. Thus, sampling risk is a function of how much evidential matter the auditor obtains during the audit.

The other aspect of risk is a function of the competence of evidential matter. It involves the possibility that applying the procedures to all details of the transactions or balances might fail to detect a material error that occurs or fail to reveal compliance deviations that would influence the auditor's evaluation of the system of internal control. This aspect is known as the *nonsampling risk,* and it is attributable to the nature of the audit procedures, the timing of the procedures, the system being examined, and the skill and care of the auditor. Controlling this nonsampling risk is very important and should be carefully considered by the auditor in determining the nature and timing of the auditing procedures.

The distinction between the two aspects of risk is recognized in section 320A.17, which states:

> The competence of evidential matter as referred to in the third standard of field work is solely a matter of auditing judgment that is not comprehended in the statistical design and evaluation of an audit sample. In a strict sense, the statistical evaluation relates only to the probability that items having certain characteristics in terms of monetary amounts,

quantities, errors, or other features of interest will be included in the sample—not the auditor's treatment of such items. Consequently, the use of statistical sampling does not directly affect the auditor's decisions as to the auditing procedures to be performed, the acceptability of the evidential matter obtained with respect to individual items in the sample, or the action which might be taken in the light of the nature and cause of particular errors.

Design of the audit program entails considering control over each aspect of audit risk—both sampling risk and nonsampling risk—in both the substantive tests as well as any compliance tests. For each type of test, the risk attributable to sampling may be considered as an additional risk over and above the nonsampling risk. As an approximation, the auditor may regard his total risk as being the *sum* of the two risks.

For example, an auditor may examine a sample of sales orders to determine whether credit sales are being approved as required. When the only evidence that an order was properly approved is the presence of an authorized signature, there is some risk that the sale was not, in fact, approved for credit even though a signature exists. The auditor's risk of incorrectly deciding that sales have been properly approved for credit is approximately the sum of the risk that credit was not approved even though a signature exists (nonsampling risk) plus the risk that the sample incorrectly indicates the appropriate signatures are present on the sales orders not included in the sample (sampling risk).

A result of both sampling and nonsampling aspects of audit risk is that the auditor can never reduce audit risk to a lower level than the nonsampling risk. Consequently, unless the audit procedures have a nonsampling risk well below a tolerable level of audit risk, neither statistical nor nonstatistical sampling will be particularly helpful.

Compliance Tests. Section 320.55 says, "The purpose of tests of compliance is to provide reasonable assurance that the accounting control procedures are being applied as described." Accounting control procedures may be divided into two categories—those that leave an audit trail of documentary evidence and those that leave no trail. Controls in the first category can be tested by using statistical sampling, while those in the second, of necessity, are tested by nonstatistical means—inquiry and observation.

Controls from either category are grouped according to the type of error or irregularity each is designed to prevent or detect. As section 320B.20 says:

> In some situations, the primary control against a particular type of error or irregularity may be provided by a single procedure or a set of related procedures; in others, auxiliary control that is overlapping or to some degree duplicative may be provided by another procedure or set of related procedures. In either situation, a set of two or more procedures necessary for a single purpose should be regarded as a single procedure. . . .

The auditor's compliance tests of the pertinent procedures are designed to ascertain whether the preliminary evaluation is warranted. For those pertinent control procedures or sets of control procedures that leave no audit trail of evidence, the auditor's professional judgment is the basic determinant of the amount and kind of evidence required to provide the "reasonable assurance." Some risk that compliance is not as good as it appears always exists, and if the potential for management override is significant, this risk may be large. The possibility that compliance is less than the auditor tentatively expects directly contributes to the risk of unwarranted reliance. Consequently, the auditor needs to be cautious in evaluating the likelihood of the occurrence of a material error or irregularity because noncompliance with those pertinent controls may leave no audit trail of evidence.

Those pertinent accounting controls or sets of controls that leave an audit trail can be tested using a statistical sample. As previously stated, using a sample to test compliance introduces an additional source of risk. For a compliance test, this is the sampling risk that by restricting the procedures to a sample of the transactions, the auditor may decide compliance is satisfactory when, in fact, were every transaction to be examined, it would be discovered that compliance is not satisfactory. The risk of unwarranted reliance is approximately equal to the sampling risk of incorrectly deciding that compliance is satisfactory plus the nonsampling risk that the procedures used by the auditor might fail to detect noncompliance. This relationship between sampling and nonsampling risks is true whether the sampling process is statistical or judgmental. Statistical procedures allow the auditor to measure and hence control the sampling risk.

While there is a need to control the sampling risk of statistical compliance tests, it should be done in the broader context of controlling all aspects of the risk of unwarranted reliance. In determining whether the auditor's reliance is warranted, the audit procedures employed may be far more important than the sample size. The relative importance of the procedures is recognized in section 320B.16, which states:

> In addition to the statistical evaluation of the quantitative significance of deviations from pertinent procedures, consideration should be given to the qualitative aspects of the deviations. These include (a) the nature and cause of errors, such as whether they are errors in principle or in application, are deliberate or unintentional, are due to misunderstanding of instructions or to careless compliance, and the like and (b) the possible relationship of errors to other phases of the audit.

Performing a thorough error analysis on each observed compliance deviation may be far more informative to the auditor than any quantitative projections that may be obtained from a sample.

The objective of a statistical compliance test is to determine the reasonableness of the auditor's assumption of satisfactory compliance that derived from preliminary evaluation of the system of internal accounting control. His assessment of the likelihood that material amounts of errors or irregularities could occur and remain undetected assumed that compliance was satisfactory. When statistical sampling is used to test compliance, the range or rates of compliance deviation that constitute "satisfactory compliance" needs to be made explicit. How does the auditor do this? There is no completely satisfactory answer, but the following remarks may be helpful.

Compliance Deviation. Following the conceptual approach described in section 320.65, the auditor identifies the set of prevention and detection controls that have been designed to prevent or detect and correct each major type of error or irregularity. Assuming that no compliance deviations occurred, the auditor might first consider the likelihood that a material amount of error *could* occur and remain undetected within the particular account balance or set of transactions. Unless this likelihood is judged to be small, the auditor would not contemplate relying on the particular controls, and no further tests for compliance would be required.

Having established the likelihood assuming no compliance deviations as a benchmark, the auditor might then assess the effect of increasing the rate of noncompliance for those controls that leave an audit trail of evidence. This step is, of course, very difficult because it involves relating compliance deviation rates to monetary error rates. How to accomplish this is a large, unresolved problem, and the suggestions made here are offered only as tentative first steps.

The rate of compliance deviation determines the *potential* for monetary error. Of course, not all instances of procedural deviations will result in monetary errors, but the potential for such error increases as the number of procedural deviations increase. The range of satisfactory compliance might correspond to those rates of compliance deviation for which the expected potential monetary error is less than a material amount. Equivalently, the auditor might define a *threshold rate* for unsatisfactory compliance as that rate at which the expected potential monetary error equals a material amount.

How can the expected potential monetary error be calculated? A suggested way is to multiply the expected size of any monetary error times the number of transactions times the rate of compliance deviation. The nature of the error or irregularity being considered may suggest the range of monetary errors that could occur with any transaction. As a most conservative value, the auditor might consider the total amount of the transaction as the possible magnitude of the error. In this case, the expected size of the monetary error corresponds to the average amount of the transactions when compliance deviations are equally likely to occur on any transaction. During review of the system, the auditor may be able to determine whether the assumption of equal distribution of compliance deviations is tenable, and, if it is not, to use a more appropriate distribution.

For example, for a particular set of procedures used to prevent and detect unauthorized purchases, the auditor might decide that the total amount of the purchase should be used in computing the expected potential monetary error. If the total purchases amount to, say, $2 million composed of 10,000 purchases averaging $200, the expected potential monetary error corresponding to a compliance deviation rate of .01 would be equal to $20,000 ($200 × 10,000 × .01).

Alternatively, the auditor might express the noncompliance rate directly as a fraction of the dollars involved in transactions. This could be accomplished by associating the dollar amount of the transaction with each instance of a compliance deviation and expressing the noncompliance rate as the ratio of the dollar amount of transactions with compliance deviations to the total amount of the transactions. For example, if $20,000 of the $2 million of purchase transactions showed compliance deviations, the rate of compliance deviation would be .01 or 20,000/2,000,000.

The range of satisfactory compliance applies to the set of accounting control procedures that affect the occurrence of a particular type of error or irregularity. Operationally this has different implications for the auditor, depending upon the relationship among the procedures constituting the set. As stated in section 320B.20, when two or more accounting control procedures are necessary for a single purpose, they should be regarded as a single procedure, and a deviation from any one of the sets should be regarded as a noncompliance occurrence. In this case, the noncompliance rate would apply to a deviation from any one of the procedures.

When two or more procedures are overlapping or duplicated to some degree, the rate of noncompliance may correspond to lack of compliance with all the procedures in the set. For example, requiring proper approval of a purchase order is a control procedure to prevent unauthorized purchases, and requiring approval for payment is a control procedure to detect any unauthorized purchase. Regarding the two procedures as a set, the auditor could regard as an instance of noncompliance any transaction that lacks both an approval for purchase and an approval for payment. However, if a transaction failed to have evidence of only one of the two required approvals, it would not be regarded as an occurrence of compliance deviation.

Obtaining a threshold rate for noncompliance corresponding to each set of controls that leave an audit trail covering a particular cause of error or irregularity allows the auditor to design the compliance tests. For those pertinent procedures tested statistically, the auditor can control the risk of deciding that the rate of compliance deviation is below the threshold rate, when, in fact, it is at the threshold rate. Combining the risks of the statistical compliance tests covering all sources of error or irregularity allows the auditor to control the statistical risk of unwarranted reliance. For example, if there are two sources of error, the risk of unwarranted reliance would be approximately equal to the *sum* of the risks. The risks are added because unsatisfactory compliance corresponds to the rate at which noncompliance is at or above the threshold rate for either of the two sets of procedures.

Statistical samples also offer the opportunity to control another type of risk, namely, the risk that the sample evidence might incorrectly indicate that the rate of compliance deviations is above the threshold level when in fact it is below. This type of mistake can lead the auditor to undertake unnecessary audit steps that only increase the cost of the engagement. Consequently, this risk is called the *risk of overauditing*. A good auditing strategy controls the risk of overauditing—especially if the cost of additional observations to test compliance is smaller than the cost of any additional observations in expanded substantive tests.

Substantive Tests. Section 320.70 points out that the general purpose of substantive tests is "to obtain evidence as to the validity and the propriety of accounting treatment of transactions and balances, or conversely, of errors or irregularities therein." As stated in section 320B.25, "The feature of audit interest in performing substantive tests of details is the monetary amount of errors that would affect the financial statements being audited." A particular substantive test of details may test either a single source of monetary error or several sources of error. For example, a substantive test may be used to determine whether there could be pricing errors in the inventory. Likewise, confirmation requests for accounts receivable may be sent to determine whether the recorded amounts exist and are accurate. Confirmation procedures are not the primary source of evidence concerning the collectibility of the amounts.

The auditor determines the specific objectives of each substantive test of details; taken as a whole, the set of substantive tests of details results in the auditor's examining the particular account balance or class of transactions for all sources of monetary error. Whether these specified objectives will be better accomplished by applying a single sample or using several samples is determined by the auditor. Regardless of the number of separate samples, the sampling risk for each can be planned to yield an acceptable combined sampling risk for the tests taken together. This planning can be done objectively when statistical sampling is used but must be done subjectively when the extent of the sample is determined judgmentally.

The procedural steps for each substantive test of details of transactions or account balances include determining

1. The number of details to examine (extent).
2. When to examine the details (timing).
3. How to select the details for examination (nature).
4. What specific audit procedures to apply to each selected detail (nature).

The design of the substantive tests for each account balance or class of transactions should consider the alternative actions that are to be taken depending upon the sample results. In turn, reaching any conclusion concerning the existence of material error involves considering all the tests of detail— those selected statistically as well as those selected judgmentally—as well as the results of analytical review procedures.

When the sample results of all the tests taken together support the conclusion that no material error exists from any tested cause, no further action is required. When the sample results of several tests taken together suggest the possibility of material misstatement, however, the auditor takes appropriate action to determine whether there is indeed a material error. It is not possible during the preliminary design phase to stipulate exactly what actions will be most appropriate in this circumstance because the nature of the errors found will obviously influence those actions. A general requirement is to examine the observed errors for possible causes and then to investigate more thoroughly any areas affected by those causes. For example, it may be learned that a substitute clerk had committed errors that were confined to a particular accounting record or to a particular period of time; in such a case, the more thorough investigation may be concentrated on the affected record or period. When a variety of errors occur and their causes cannot be confined to particular areas, the auditor may consider extending the tests of details to obtain more corroborating evidence to determine whether an adjustment is required.

Application and Refinement

Proper execution of the statistical procedures is necessary to insure the statistical validity of the results. Substituting another item for a specified sample item, not consistently using the definition of a sample occurrence, or using an inappropriate statistical technique are some examples of faulty execution. To avoid mistakes of this type, the audit personnel conducting the tests should be appropriately trained and their work should be reviewed. Chapter 10 discusses this important area.

Evaluating the evidential matter when statistical tests are employed involves more than a statistical evaluation. Projecting the sample results to the population is certainly necessary, but the evaluation of the sample results does not stop here. Procedural deviations that are found should be examined to determine their probable cause, particularly in terms of whether they are inadvertent or deliberate. Likewise, determining the probable cause of any monetary errors is necessary, especially when the sample results indicate the possible presence of a material amount of error. This, of course, is just what the auditor would do if his sample was wholly judgmental; the fact that he uses statistical sampling techniques in no way lessens his responsibility to use sound audit procedures.

Auditing is not static. During the course of an audit, the auditor may find the situation differs from what he judged to be the case when the audit program was designed. When this occurs, the auditor must revise his audit program as required. For example, when pertinent accounting controls are found not to be operating as described, the auditor needs to revise the planned substantive tests—changing their nature, extent, or timing. The particular change adopted should reflect his overall desire to keep the audit risk at a tolerable level.

Developing an Audit Strategy

Auditing may be viewed as a complex decision problem. The auditor decides what to observe, when to observe it, and how much to observe. The accounting system itself is often extremely complex, and, consequently, the auditing procedures must cope with that complexity.

Developing an appropriate audit strategy involves balancing audit risk on the one hand with time and cost considerations on the other. This is expressed in section 330.09, which states:

> The amount and kinds of evidential matter required to support an informed opinion are matters for the auditor to determine in the exercise of his professional judgment after a careful study of the circumstances in the particular case. In making such decisions, he should consider the nature of the item under examination; the materiality of possible errors and irregularities; the degree of risk involved, which is dependent on the adequacy of the internal control and susceptibility of the given item to conversion, manipulation, or misstatement; and the kinds and competence of evidential matter available.

Section 330.12 states:

> An auditor typically works within economic limits; his opinion, to be economically useful, must be formulated within a reasonable length of time and at reasonable cost. The auditor must decide, again exercising professional judgment, whether the evidential matter available to him within the limits of time and cost is sufficient to justify formulation and expression of an opinion.

As discussed in the previous section, the auditor may control the risk of failing to detect a material error by controlling both the risk that substantive tests taken together fail to detect a material amount of monetary error and any risk that an unwarranted degree of reliance is placed on the system of internal accounting control. In addition to their sampling aspects, each of these risks also has nonsampling aspects.

The nonsampling aspect is the most difficult to assess because it encompasses so many unmeasurable factors. What is the likelihood that a given audit procedure (for example, direct communication with debtors) or set of procedures used in a particular organization will fail to detect a material amount of monetary error? Does that likelihood differ between monetary error of overstatement and understatement? What is the likelihood that the audit personnel performing those procedures possess the skill necessary to recognize a monetary error? Notwithstanding the difficulty of coping with these and other questions, the auditor needs to address the question of how to control the nonsampling aspects of risk in developing a good audit strategy.

Thus, an objective of a good audit strategy is to reduce the audit risk to a tolerable level in the most effective and economical way possible within any time limits that are imposed. Ideally, the auditor would like to select the optimum strategy but practically selects a strategy that is satisfactory and is the best available in the circumstances. The chief reason for the gap between what the auditor would like and what can be achieved is the difficulty of measuring the total audit risk associated with any program, particularly the nonsampling aspect.

Controlling Audit Risk

A complete solution to the complex problem of controlling audit risk is not available, but improvement of the present methods of coping with this problem is possible. A suggested step in this direction is to give more attention to those facets of audit risk that are susceptible to objective measurement. As previously described, the audit risk of any test of details is composed of both sampling and nonsampling aspects. Moreover, as an approximation, the sampling risk may be added to the nonsampling risk to obtain the audit risk for a particular test.[5] This makes it possible for the auditor to consider the sampling risk separately from the nonsampling risk when designing the audit program.

Controlling the sampling risk does involve considering the nonsampling aspects of other tests or audit procedures directed at the same sources of errors or irregularities. For example, when a statistical test of details is combined with a nonstatistical analytical review procedure, the resulting audit risk of the auditor's failing to detect a material error equals the product of the two audit risks. Expressing the audit risk of the statistical test as the sum of its sampling and nonsampling aspects, the combined audit risk equals the sampling risk times the likelihood that the analytical review procedure fails to detect a material error plus the nonsampling risk times the same likelihood. The sampling portion of the combined audit risk is thus equal to the sampling risk of the test of details times the subjective risk of the analytical review procedure. For example, if the sampling risk of a substantive test of an inventory's prices, quantities, and extensions were .30, and the analytical review procedures were judged to have a 50 percent likelihood of not discovering a material amount of error in the inventory amount, the sampling portion of the combined risk would be $.30 \times .50 = .15$.

Determining a tolerable level of audit risk for tests of details pertaining to each account balance or class of transactions should be done in the context of

1. The degree of planned reliance on the pertinent internal accounting control procedures.
2. The procedures designed to appraise the overall reasonableness of the recorded amounts, such as analytical review procedures.

The maximum degree of planned reliance is based on the auditor's review and preliminary evaluation of the system, assuming satisfactory compliance. The auditor may decide to use less than the maximum degree if the total cost of the audit program—including any required compliance tests—can be lowered.

A major question concerns expressing the degree of planned reliance. Section 320B.35 suggests using a numerical scale (C) wherein the planned reliance is expressed as a number between zero and one. Because of the limitations discussed in section 320.34, C would always be less than one. Ideally, C would be determined as the likelihood that the set of pertinent accounting controls would prevent or detect a material amount of error. In practice, the auditor may either assign a numerical value to C or use a scoring scheme such as suggested by Elliott and Rogers [10]. Their scheme, or something similar, may be a useful method for expressing the degree of reliance.

Section 320B.35 goes on to suggest that the audit risk for substantive tests should be determined so that the product of it times $(1 - C)$ equals a tolerable combined risk level (labeled $(1 - R)$). Following this suggestion entails designing the audit pro-

5. The reason for this is as follows: The probability that a sample detects a material error equals the probability that the procedure applied to every detail would detect a material error multiplied by the probability that the sample yields the same result as applying the procedure to every detail. The probability that a sample fails to detect a material error is the complement of this product, and this can be reasonably approximated by the sum of the nonsampling risk and the sampling risk. For example, suppose the procedure has a likelihood of .85 of detecting a material error, and the statistical test has a reliability of .95. Then the likelihood that the procedure fails to detect a material error equals $1 - (.85)(.95) = .1925$ which is approximately .15 plus .05.

gram so that for each account balance or class of transactions, the product of the likelihood that the accounting control system could allow a material amount of error to occur and remain undetected times the auditor's risk that substantive tests taken as a whole could miss a material error is equal to a constant—the tolerable combined risk level. For example, if the auditor were to assign numerical values to the risks, the following results for accounts receivable might be found:

Internal accounting control:			
Assign a likelihood that a material error could occur and remain undetected			.40 $(1 - C)$
Tests of details:			
Assign a likelihood of nonsampling risk	.10		
Add sampling risk for statistical tests (or equivalents likelihood for judgmental sample)	.20		
Conservative approximation of audit risk		.30	
Analytical review:			
Assign likelihood of failing to detect a material error		.50	
Audit risk for substantive tests (.30 × .50)			.15
Combined risk level (.40 × .15)			.06 $(1 - R)$

Notice that the nonsampling portion of the combined risk level is .02 (.4 × .1 × .5), while the sampling portion is .04 (.4 × .2 × .5). The discussion of audit program planning in chapter 7 will be limited to the sampling portion of the combined risk, which will be labeled $(1 - RS)$. The importance of keeping the nonsampling portion of the combined risk at a tolerable level is addressed in SAS no. 1 and need not be repeated here.

Evaluating Trade-offs

One of the important decisions the auditor makes is the degree of his planned reliance on internal accounting control. If the auditor decides to rely on the system to the maximum extent possible, the nature, timing, and extent of his substantive tests are planned to yield a tolerable combined risk level. Using this maximum planned reliance, however, the auditor must conduct compliance tests of the pertinent controls to obtain satisfaction that the system is operating in the prescribed manner. The direct consequences of this decision are

1. The risk that the planned reliance is unwarranted.
2. The cost of testing compliance with the pertinent procedures.

On the other hand, the auditor may decide not to rely on the internal accounting controls but, instead, obtain the same combined risk level solely through substantive procedures including tests of details plus analytic review. This decision avoids the risk of unwarranted reliance and the cost of compliance tests, but may substantially increase the cost of the substantive procedures.

Which is the better decision? Answering this entails considering both the audit risk and the total cost of the alternatives. For example, suppose the auditor sets the tolerable combined risk ($1 - R$) at .05, and determines that the maximum degree of reliance (C) is represented by .80. If the auditor uses this maximum degree, the audit risk for substantive tests would be .25($(1 - .80) \times .25 = .05$). On the other hand, if the auditor uses minimal reliance, the audit risk for substantive tests would be .05 ($(1 - 0) \times .05 = .05$). The .20 difference between the two risk levels (.25 − .05) multiplied times the risk of unwarranted reliance is a measure of the additional risk caused by uncertainty concerning the efficacy of the system of internal control. If, in this example, the risk of unwarranted reliance is .10, the additional risk is .02 ($.10 \times .20$) if the auditor relies on the controls and sets the tolerable audit risk for substantive tests at .25.

One criterion to use in deciding the appropriate level of planned reliance would be to select that alternative with the lower audit cost subject to the condition that the additional risk should not exceed some stipulated level, say .01. Using this criterion in the above example would mean the auditor should decide between using minimal reliance with an audit risk of .05 for substantive tests and using the maximum reliance with an audit risk of .25 together with a risk of unwarranted reliance equal to .05 ($(.25 - .05) \times .05 = .01$). The auditor would select the one having the smaller cost.

To obtain the relevant cost comparisons of the two alternatives, the auditor needs to take into account the possibility that compliance tests might result in placing less than the planned reliance on the system of internal accounting control even when compliance is satisfactory. This may occur when the sample results incorrectly indicate that the compliance deviations could exceed the threshold rate.

The total cost under minimal reliance is just the cost of performing the substantive tests at the sampling risk corresponding to minimal reliance. Thus, this total cost should include any extra cost because of the choice of timing or nature of the procedures, in addition to the extent of the procedures.

This is to be compared to the total cost under the maximum possible reliance, including the cost of conducting the compliance tests of pertinent procedures. To this total cost should be added the difference in cost between conducting the substantive tests at minimal reliance and the maximum reliance times the probability of over-auditing. This additional factor corresponds to the additional expected cost incurred when the compliance tests result in incorrectly deciding that compliance is unsatisfactory, and, consequently, the auditor decides to use the minimal reliance in conducting substantive tests.

Evaluating such trade-offs is one important aspect of generating a good audit strategy. Measuring the sampling risk permits improvement of the ability to make decisions pertaining to the extent of substantive tests of detail. At the same time, the nonsampling aspects of risk must weigh heavily in the auditor's determination of the nature and timing of audit procedures. The discussion here illustrates how interrelated these two aspects are. Consequently, while statistical sampling allows the auditor to improve the effectiveness of the audit program, controlling the nonsampling aspects of risk may be of equal or greater importance.

Summary

A description of those portions of the audit process affected by the auditor's decision to use statistical sampling reveals the major contribution that statistical sampling offers—the opportunity to measure, and hence control, the sampling portion of audit risk.

The auditor can use statistical samples both for compliance tests and substantive tests. For compliance tests, use of statistical sampling allows the auditor to control the risk of unwarranted reliance. This is the risk of deciding that the rate of compliance deviation does not exceed some threshold rate when, in fact, the rate of compliance deviation is unsatisfactory (equals the threshold rate).

For substantive tests, statistical samples permit the auditor to control the risk that procedures would fail to reveal the existence of a material amount of monetary error.

Planning all the tests pertaining to a particular account balance or class of transactions together allows the auditor to control the total sampling risk. In addition, there is an opportunity to evaluate the relative merits of several alternative plans and select the one that is preferred.

2

Basic Statistical Concepts

This chapter introduces the basic statistical concepts important to the auditor. The objective is to present those concepts required for understanding and using statistical techniques to select a sample and project the results to the whole from which the sample was selected.

Probability

The foundation of all statistical sampling is probability theory. Two distinct interpretations of the word *probability* are used when this theory is applied to practical problems. One meaning is that of long-run frequency. If the probability of choosing a specified item is .01, then repeated choices made under the same circumstances would result in the item's being selected in about one percent of the cases. For example, in one thousand choices, about ten would contain the specified item. This frequency interpretation is the basis of statistical sampling theory and is used throughout most of this book. When using this meaning of probability, the auditor is stating that repeated application of a procedure would produce the same result at a stated frequency.

Another interpretation of probability is as a subjective measure of belief. In that usage, the meaning of the statement "the probability of a material overstatement is .01" is that the person believes the existence of a material overstatement rates .01 on a scale of zero to one (1.00). It also suggests that the person would be willing to give odds of 99 to 1 against the existence of material overstatement. Odds are simply the probability of one outcome divided by the probability of another outcome. In this illustration, the probability of no material misstatement is .99 (1.0 − .01) and the odds are, therefore, .99/.01 or 99 to 1.

Successfully integrating statistical sampling into the audit process entails using the subjective interpretation of probability to express the auditor's judgment concerning some uncertain events. Making audit decisions regarding the nature, extent, and timing of audit tests (what, when, and how much to examine) can be based on blending the frequency interpretation with the subjective interpretation to express the auditor's risk.

Using a sampling method that gives each possible sample a known probability (frequency) of being selected, permits the auditor to make probabilistic inferences about the whole population. To accomplish that, it is necessary to specify the *population,* the *sampling unit,* the *frame,* the *characteristic* of interest, and the *selection method.* The *statistical* technique to be applied must be specified. Commonly used techniques are discussed in chapters 4, 5, and 6.

Population

The population is the aggregate of accounting entries about which information is desired.[1] The particular population is defined to serve the auditor's objectives. The only stipulation is that the definition should permit anyone to tell whether an item either belongs or does not belong to the population. For example, if the population is defined as being *all* accounts receivable as of October 1, accounts with zero balances, credit balances, and unrecorded balances are included. If these accounts are not to be included, the population must be defined in a more limited way. For example, defining the population as all accounts receivable balances appearing on a trial balance of the subsidiary records would exclude those accounts with zero balances as well as unrecorded balances. Similarly, restricting the population to accounts with debit balances would exclude any accounts with zero or credit balances.

A general description of a population is not sufficient for selecting a statistical sample. Operationally, the auditor needs to specify the individual members of the population, called the *sampling units;* the physical representation of the population, called the *frame;* and the particular *characteristic* or characteristics of each sampling unit that is of audit interest.

Sampling Unit

The individual members of the population are called the *sampling units.* For example, the sampling units for an accounts receivable population might be the individual accounts. The client's record-keeping system, the particular statistical technique to be applied, and the audit objective will help the auditor determine the most appropriate sampling unit to use. Individual accounts are a natural choice, but requesting confirmation of account balances may be impracticable. The details of accounts receivable may consist of copies of uncollected sales invoices, and it may not be practical to summarize them by customer. Even when balances are available, to obtain confirmation from government agencies or certain other customers may require limiting requests to individual contracts or sales invoices. For such situations, using uncollected line items or invoices as sampling units might be a better choice.

Similar considerations arise when the auditor defines the sampling unit in a population of transactions covering a particular period. For example, the auditor may use either vouchers or checks as the sampling unit. Because a single check may pay

---

1. Other terms used for this concept are *universe* and *field.* See Arkin [4] pp. 20–25 for a more complete discussion.

several vouchers, auditing all the details supporting a check may take more effort than auditing all the details supporting a voucher.[2]

The auditor should also consider his audit objectives. Several objectives can be satisfied using a single sample, provided the sampling unit is appropriately selected. For example, selecting a voucher as the sampling unit enables the auditor to satisfy his objectives with respect to recorded purchases and the associated disbursements.

Finally, selecting a sampling unit often affects the efficiency of the statistical technique.[3] Populations composed of small sampling units may be more homogeneous than large sampling units. For example, individual credit sales may vary far less than customer balances. The advantage of using the smaller unit is that the required sample size may be smaller.

Frame

The frame is a listing or other physical representation of the sampling units. If individual accounts receivable are the sampling units, the frame may be defined either as the trial balance (a listing) of the subsidiary records or as the ledger cards (a physical representation) constituting the subsidiary records.

Often a variety of potential frames is available, but the one requirement of a frame is that it be a *complete* representation of all the sampling units constituting the population. Among the alternative frames satisfying this requirement, the auditor chooses the one that is most practical for accomplishing the audit objective. For example, in examining inventories, alternative frames might be (1) a listing showing quantities, prices, and extensions prepared as the result of either a physical inventory or an update of perpetual records; (2) perpetual records showing quantities but not prices; or (3) a map of physical locations or an index of tag numbers assigned to them.

Of these alternatives, the listing would be the most practical frame when the objectives are to determine whether all items in the record do, in fact, exist and whether recorded quantities, prices, and extensions are correct. On the other hand, if the listing is not available until too late in the audit, either of the other frames is acceptable, provided the auditor can establish that it is complete.

Since completeness is the principal requirement, and his conclusions pertain only to the set of sampling units included in the frame, the auditor must be satisfied that the frame does contain all the sampling units corresponding to the defined population. Continuing the inventory example, the auditor could validate a listing by (1) footing it to see that its total agrees with the general ledger and (2) tracing items from the physical locations to see that they are represented on the listing. Similarly, to determine that the index of tags assigned to inventory locations is complete, the auditor would need to determine that a tag had been issued to each listed inventory item.

Frames should be complete, but they may contain units that do not belong to the population of sampling units. For example, a listing of accounts receivable may contain inactive accounts that are not part of the defined population to be sampled. This presents no statistical difficulty. In selecting the sample from the frame, any unit selected that does not belong to the population is simply ignored for statistical purposes. For audit purposes, examining selected items that do not belong to the popula-

2. When some form of *two-stage sampling* can be employed, such additional effort can be reduced.
3. See Appendix 1 for a definition of *statistical efficiency*.

tion may be very important. For example, a check number corresponding to a voided check may be examined to see that it was properly voided even though it is not part of the population of disbursements.

Characteristic

The purpose of sampling is to infer something about a characteristic of the population. To make an inference about a specified population characteristic, a corresponding characteristic for each sampling unit should be defined. One typical population characteristic is the total audited dollar amount. To permit inferences about this population characteristic, it must be possible to determine an audited dollar amount for each sampling unit.

The auditor may specify one or more characteristics for each sampling unit. For example, the recorded dollar amount, the audited dollar amount, and the difference amount (audited amount minus recorded amount) are three characteristics that may be associated with a sampling unit. Each of these characteristics is an example of what is called a *variable.* A variable is any quantitative characteristic associated with a sampling unit. In auditing, the variables encountered are expressed typically in dollars.

Another type of characteristic that may be associated with the sampling units is known as an *attribute.* An attribute is a qualitative characteristic, such as a procedural deviation, that either does or does not exist within a population. Consequently, each sampling unit either possesses or fails to possess the attribute. In this sense, attributes are measured in terms of the *number* of occurrences, and the corresponding population characteristic is typically the *percent* of the population units that possess the attribute.

In auditing applications, considerable care must be exercised in defining the attribute. For example, if the sampling unit is a recorded check and a defined attribute is lack of a properly authorized purchase order, there may be checks for which no purchase order is required. Consequently the lack of a purchase order would not necessarily indicate a deviation from internal control procedures. One way to avoid this is to define the population so as to eliminate all checks that do not require purchase orders. Another way is to define the attribute as lack of proper authorization, with the different ways a check could be properly authorized explicitly noted. The auditor selects the alternative that permits accomplishing the test objective.

A common practice is to define several attributes for each sampling unit. Thus, the auditor might define the several attributes to be particular types of deviations in such functions as authorization, pricing, approval, and so forth. Each attribute, however, is separately evaluated, and this may mean having a slightly different population of sampling units for each attribute. For example, all recorded checks could be the appropriate population for testing accounting distribution, and only those recorded checks for purchases requiring purchase orders could be the appropriate population for testing the existence of a properly authorized purchase order.

Finally, the nature of the sampling unit influences what constitutes an occurrence of the defined attribute. In the case of one check that pays several invoices, the absence of either one or several purchase orders would count as a single occurrence.

Selection Method

The validity of any statistical inference depends on the manner in which the sampling units are selected. Only those methods that result in a known probability of selecting each sampling unit in the population permit valid statistical inferences. This means that attempts to select randomly do not necessarily meet the criterion unless rigorous techniques are applied. Likewise, when the sampling units are selected judgmentally, valid statistical inferences are not possible.

The selection methods described in this chapter include unrestricted random sampling, systematic sampling, and sampling with probability proportional to size.

Unrestricted Random Sampling

An unrestricted random sample of size *n* has two defining properties: (1) each sampling unit has an equal probability of selection, and (2) each possible group of *n* units has an equal probability of selection. A given sampling unit may be selected either with or without replacement. *Selection with replacement* means that each unit in the sample is selected from the entire frame so that a particular unit may be selected more than once. *Selection without replacement* means that once a unit is selected it is effectively removed from the frame; thus, the sample will always consist of *n* different sampling units. Sampling without replacement is ordinarily used in auditing.

To select an unrestricted random sample without replacement, the auditor may use either a random number table or a computer program that generates random numbers. To use a random number table, the auditor needs first to establish a correspondence between the digits in the table and the sampling units.[4] If the sampling units have identifying numbers, that number could be used. If the sampling units are listed on several pages, the page and line numbers could form the basis of the correspondence.

For example, if the frame consists of a listing of 300 pages with no more than 20 sampling units per page, correspondence could be established by using a five-digit number. The first three digits would correspond to the page number and the last two would correspond to the line number. The number 00101 corresponds to page 1, line 1, and the number 30020 corresponds to page 300 line 20. When some pages have fewer than 20 units, no bias is introduced if the auditor ignores numbers corresponding to blank positions on those pages.

The correspondence described above is valid, but it involves discarding many five-digit numbers. To overcome this inefficiency, a scheme of multiple correspondence may be established whereby each sampling unit corresponds to more than one random number. In devising such multiple correspondence, care should be taken to assure that each sampling unit corresponds to the same number of random numbers.

For example, 30101 and 60101 could also identify page 1, line 1, and 60020 and 90020 could identify page 300, line 20. This scheme identifies each page and line number by three five-digit random numbers and reduces the frequency of unusable random numbers. If desired, the number of random numbers assigned to each page and line number could be increased by assigning to page 1, line 1, the following.

00101	30101	60101
00121	30121	60121
00141	30141	60141
00161	30161	60161
00181	30181	60181

4. Arkin [4] contains a table of 105,000 random digits.

If this correspondence is used, the twentieth item on any page has a correspondence pattern represented by page 300, line 20, as follows.

30020	60020	90020
30040	60040	90040
30060	60060	90060
30080	60080	90080
30000	60000	90000

Notice that the bottom row uses 30000; this is required to give the twentieth item an equal opportunity of selection.

In deciding on the correspondence scheme, two important requirements are to (1) make sure the scheme gives each sampling unit an equal chance of being selected, and (2) avoid complexities that might result in misapplication in the field. For example, if there are 338 pages and 21 lines per page, the following correspondence pattern might be used for page 1, line 1: 00101, 00131, 00161, 40101, 40131, 40161. In devising this scheme, the number of pages was taken as 400 and the number of lines as 30 to make the scheme simpler.

Systematic Sampling

Systematic sampling is a convenient method of selection if properly used. Although the method is easy to use, it may give invalid results under certain circumstances (as discussed below). In one form of systematic selection, the auditor selects the first sampling unit at random from among the first k sampling units and thereafter selects every kth sampling unit from the entire frame. To determine the value of k, divide the population size (N) by the desired sample size (n) and then set k equal to the closest integer smaller than N/n. Following this procedure insures that the sample will be at least n. For example, if a sample of 90 units is required from a population of 10,000, $k = 111$ following the suggested procedure because $10{,}000/90 = 111.11$. The first sampling unit is selected at random from the first 111 units. Then, the sampling units are counted (beginning with the number one), and each time the count reaches 111, the corresponding unit is selected. To demonstrate the results of this method, suppose that the units are consecutively numbered and that the first unit selected at random is 85. Then the other units in the sample correspond to numbers: 196 $(85 + 111)$, 307 $(196 + 111)$, 418 $(307 + 111)$, and so forth. (Even when the units are not consecutively numbered, counting the sampling interval in this way effectively assigns consecutive numbers to them and thereby arrives at exactly the same units.)

The selection procedure needs to be continued throughout the frame. In the above example, this means that if the first sampling unit selected at random is from one to ten, there will be 91 units in the sample. To stop the selection after 90 units would bias the results since the 10,000th population unit would have no chance of being selected. Another point to remember is that no substitution is permitted. If, in the example, unit 196 corresponds to an unusable item, the auditor must not substitute unit 195, unit 197, or any other unit—the sample size is simply one less.

A systematic sample is certainly a probability sample; in certain circumstances, it produces results that are essentially equivalent to unrestricted random sampling.[5]

5. See Cochran [5], p. 210, for conditions under which systematic sampling is superior to unrestricted random sampling.

The required circumstances are that the sampling units be in random order with respect to the characteristic being measured. For example, if the auditor's objective in a price test of inventory items is to discover errors, a systematic sample would produce results equivalent to an unrestricted random sample only if the errors are randomly distributed among the inventory items in the frame.

It is undoubtedly reasonable to assume in many situations that the characteristic selected by the auditor appears in random order within the frame. However, in making that assumption, the auditor introduces some risk and must decide whether this risk is worth the convenience of systematic sampling.

One frequently used method that reduces the dependence on the assumed randomness is the use of several random starts instead of a single start. In the example of the sample of 90 from a population of 10,000, the auditor might use ten random starts and a sampling interval of 1111. The ten random starts yield ten subsamples, each containing nine sampling units. To avoid difficulties in using this method, it would probably be better to use a skip interval of 1100. This would result in ten subsamples some of which could contain ten units. Evaluating the sample results may then be done using the techniques of *replicated sampling*.[6]

Sampling Proportional to Size

Another method of obtaining a probability sample is to select the sampling units with probabilities proportional to the size of the unit. This method, abbreviated pps, can be very useful in auditing. The measure of size used is often the recorded amounts. This gives units with large recorded amounts more opportunity to be selected than units with small recorded amounts, and, typically, the auditor is more interested in examining the larger units. Because sampling units having a recorded amount of zero cannot be selected, they must be treated separately, as should sampling units with negative recorded amounts.[7]

Several methods may be used to select a pps sample. When no negative recorded amounts exist, one method of choosing a sample of size n is as follows:

1. Select n random numbers from a random number table. Each random number should contain as many digits as are contained in the *total* recorded amount.
2. Arrange the selected random numbers in ascending order.
3. Prepare a listing of the accumulated recorded amounts for the population of sampling units.
4. Find the location in the listing of the selected random numbers and take the corresponding sampling unit into the sample.

To illustrate, if the recorded amount is $5,284 and a sample size of three is desired, the auditor selects three random numbers, each with four digits not greater than 5284. Suppose these are 4717, 4149, and 0740, as read from the table, and

6. Deming [7] and Arkin [4] describe the procedures that may be used to do this. In using this method (sometimes called replicated sampling), the auditor must decide on the number of subsamples. In general, the more subsamples the better, but some authors have suggested using at least ten (Arkin [4]).
7. Negative recorded amounts refer to credit balances among asset accounts or to debit balances among liability accounts. Reversing transactions such as NSF checks deducted from cash receipts or sales returns deducted from sales would also be negative.

become 0740, 4149, and 4717 in ascending order. This says that the sample should consist of the sampling units containing the 740th accumulated dollar, the 4149th accumulated dollar and the 4717th accumulated dollar. Suppose the population consists of ten items, as follows:

Unit No.	Recorded Amount	Accumulated Amount
1	$ 448	$ 448
2	641	1089
3	167	1256
4	342	1598
5	1066	2664
6	789	3453
7	347	3800
8	578	4378
9	728	5106
10	178	5284

The random number 0740 occurs in unit 2, the number 4149 occurs in unit 8, and 4717 occurs in unit 9. So, the sample of three consists of units 2, 8, and 9 with the respective recorded amounts of $641, $578, and $728.

Had the random numbers been 1629, 2532, and 4178, unit 5 would be selected twice, since both 1629 and 2532 correspond to it. The illustrated method thus permits a sampling unit to be selected more than once—so the sampling is effectively with replacement. Because the evaluation techniques take this possibility into account, such duplication as may occasionally occur does not present any difficulty to the auditor.

There is an alternative scheme that does not require accumulating the recorded amounts.[8] This scheme consists of selecting a random number between one and N, where N represents the total number of sampling units and the unit corresponding to that random number is then selected as a provisional member of the sample. A second random number is then selected between one and the *largest* individual recorded amount. If this second random number does not exceed the recorded amount of the provisional unit, that unit is taken into the sample; otherwise, it is not. This procedure is repeated until the requisite number of sampling units has been selected.

The advantage of this alternative is that it does not require the cumulation of the recorded amounts. The disadvantage is that a relatively large number of trials will have to be performed before a sample of given size is selected. If the sample size is large, the effort in selecting the sample in this way may outweigh the effort of cumulating the recorded amounts.

A third method is to use a form of systematic selection. One way to do this is to specify a sampling interval of k dollars. When the desired sample size is n, the sampling interval k is set equal to the total recorded amount (Y) divided by n. Whenever this quotient is not an integer, k is set equal to the largest integer smaller than

8. See Raj [18], p. 48.

Y/n. If the total recorded amount (Y) is $991,269 and the desired sample size (n) is 67 items, for example, the sampling interval would be rounded down from $14,795.06 to $14,795. The procedure is as follows:

1. Select a random number between one and k. Call it g.
2. Cumulate the recorded amounts of the sampling units in the frame whose recorded amounts are less than k until g is first equaled or exceeded. The corresponding sampling unit is taken into the sample.
3. Continue cumulating the recorded amounts that are less than k. Take into the sample those sampling units that cause the cumulated amounts to equal or exceed $k + g$, $2k + g$, $3k + g$ . . . .
4. Any sampling unit whose recorded amount equals or exceeds k is automatically a part of the sample, and the recorded amounts of such units do not enter the accumulation process.

Applying this procedure to the previous example to obtain a sample of three, the auditor determines the skip interval of $1761 by dividing the total recorded amount ($5284) by three. The starting number of 0354 was selected from the random number table.[9] The sample numbers are 0354, 2115, and 3876. This corresponds to sample item numbers 1, 5, and 8 with recorded amounts of $448, $1066, and $578, respectively.

Following this procedure, the final sample will include (1) those sampling units with recorded amounts less than k corresponding to g, $g + k$, $g + 2k$, and so forth, and (2) all sampling units with recorded amounts equal to or greater than k. The resulting sampling can be treated as a pps sample of all sampling units with a recorded amount less than k provided the characteristic of interest (often the difference between the audited and recorded amounts) is unrelated to its position within the frame (that is, the frame is in random order with respect to the characteristic).

This is essentially the same condition that was cited previously with systematic selection. As before, it is undoubtedly a reasonable assumption in many situations. The auditor must decide whether the risk that the assumption is not true is worth the convenience of this or a similar selection process.[10]

Population Distribution

The key concept underlying any statistical technique is the *sampling distribution*. The sampling distribution describes the set of all possible sample outcomes and is closely related to the *population distribution*. The characteristic of interest to the auditor generally does not have the same value for each sampling unit. There is usually a set of possible values for the sampling units; the population distribution describes the fraction of the sampling units that possess each possible value of the characteristic. The importance of the population distribution to the auditor is twofold:

9. This number was read from the Table of 105,000 Random Decimal Digits, Statement 4914, Interstate Commerce Commission, May, 1949, presented in Arkin [4], beginning on line 104, column 8 and reading down until the first four-digit number equal to or less than 1761 was encountered.
10. The Auditape system of Haskins & Sells uses a somewhat different selection technique that also results in a type of systematic pps sample but with more randomization.

(1) if he knew the population distribution of the characteristic, he would be fully informed and would not require a sample, and (2) the population distribution affects the results of the sampling process (the sampling distribution).

The population distribution may be described by a *frequency distribution.* A frequency distribution shows the fraction of the sampling units that possess each possible value of the characteristic. A frequency distribution may be portrayed either by a table or a graph; in either case, the first step is to divide the range of possible values of the characteristic into class intervals.[11] Then, the frequency with which the characteristic's values fall into each class interval is expressed as a fraction. For example, if the population is 500 checks with amounts ranging from $10 to $450 and totalling $67,500, a frequency distribution in tabular form would be as follows.

Class Interval	Number of Units	Frequency	Cumulative Frequency
1– 50	175	.35	.35
51–100	75	.15	.50
101–150	60	.12	.62
151–200	45	.09	.71
201–250	45	.09	.80
251–300	40	.08	.88
301–350	35	.07	.95
351–400	20	.04	.99
401–450	5	.01	1.00
	500	1.00	

In graphical form the same frequency distribution would appear as follows.

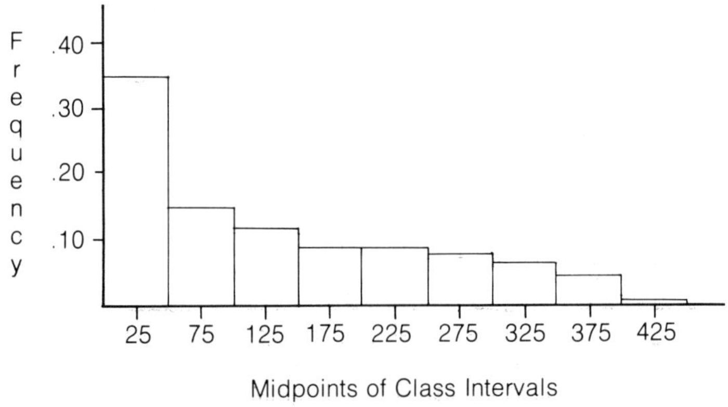

Figure 2.1

In this illustration, the characteristic of the sampling unit was a variable—the recorded dollar amount of the check. The frequency distribution describes the fraction of the sampling units falling into each class interval of recorded check amounts.

11. There are conventions for choosing the number and size of the intervals. Descriptions of these conventions may be found in [22].

Examining either the table or the graph shows that .35 of the checks were for recorded amounts from $1 to $50, that .62 of the checks were for recorded amounts not exceeding $150 and that only .05 of the checks were for recorded amounts exceeding $350. Therefore, from a frequency distribution, the auditor can obtain complete information about how the characteristic varies among the sampling units.

Instead of the recorded check amount being the characteristic, the auditor might choose to define the characteristic as the difference between the audited amount and the recorded amount. Were he to audit each check in the population, these differences might have a frequency distribution such as the following.

Class Interval	Number of Units	Frequency	Cumulative Frequency
$−4.49 − −3.50	0	.00	.00
−3.49 − −2.50	5	.01	.01
−2.49 − −1.50	5	.01	.02
−1.49 − − .50	10	.02	.04
− .49 − + .50	450	.90	.94
+ .51 − +1.50	15	.03	.97
+1.51 − +2.50	5	.01	.98
+2.51 − +3.50	5	.01	.99
+3.51 − +4.50	5	.01	1.00
+4.51 − +5.50	0	.00	
	500	1.00	

In graphical form this frequency distribution would appear as follows.

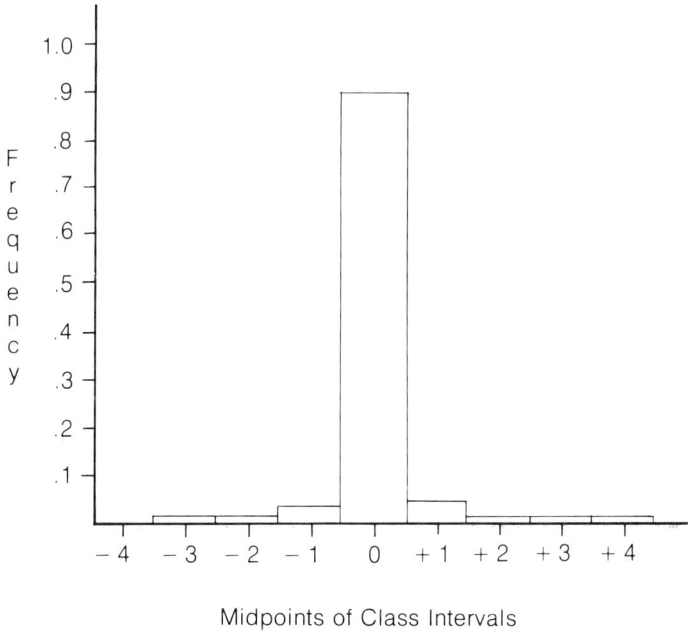

Figure 2.2

The contrast between the two illustrated distributions of recorded check amounts and difference amounts is striking. The distribution of difference amounts is heavily concentrated about zero (.90 of the checks have differences of less than $.50), and is nearly symmetric—that is, the frequency and amounts of positive differences are about the same as the frequency and amounts of the negative differences. The distribution of recorded check amounts, on the other hand, does not show any such symmetry—the technical term for the shape is *skewed.*[12] It is quite common for the distribution of recorded amounts in accounting populations to exhibit skewness that is even greater than in the illustration.

As an illustration of an attribute, suppose each check is classified according to whether proper supporting documents exist. The frequency distribution of this attribute is described by knowing what fraction of the sampling units *fail* to have proper support. To describe the frequency distribution in graphical form, an occurrence (absence of proper support) is coded as "1" and a nonoccurrence (presence of proper support) is coded as "0." The graph corresponding to an occurrence rate of .08 is shown as follows.

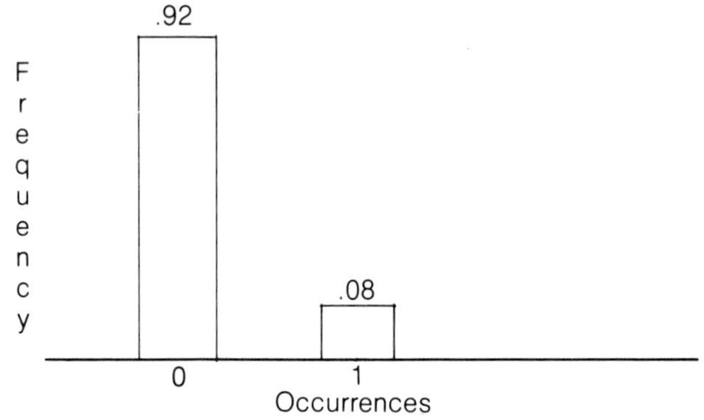

Figure 2.3

These examples illustrate that the concept of the population distribution applies to any characteristic the auditor selects, whether it is a variable (a dollar amount) or an attribute (a compliance deviation). In each case, the population distribution completely describes how the defined characteristic varies among the sampling units. Determining such a distribution is often impractical (in some cases a complete audit would be required), and, consequently, various summary measures are widely used to describe the population. Some of these are called *measures of location* and others are called *measures of dispersion.* In statistical sampling the most important measure of location is the *population mean* and the most important measure of dispersion is the *population standard deviation.*

Population Mean

The population mean locates the center of the frequency distribution. It is computed by adding the characteristics of each sampling unit and then dividing by the total number of sampling units. For the previous illustration of the population of 500 checks totalling $67,500, the population mean is $135. Capital letters are used in this

12. See Appendix 1 for an explanation of the term.

book to refer to values of population characteristics. Thus, if Y refers to the total recorded amount of the checks, and N the total number of checks, the calculation of the population mean ($\bar{Y}$) may be represented by the following formula:

$$\bar{Y} = \frac{Y}{N}.$$

In turn, $Y = \sum Y_j$, where Y_j refers to the recorded amount of the jth sampling unit and capital sigma ($\sum$) indicates that all amounts should be summed.[13]

When the characteristic is an attribute (for example, the check lacks proper support), the population mean is the proportion of sampling units lacking proper support. This proportion is denoted by P. In the case of attributes, the population mean is called the population occurrence rate and is often expressed as a percent rather than as a decimal fraction.

Population Standard Deviation

The population standard deviation measures the dispersion of the distribution about the population mean. The larger the standard deviation, the more widely dispersed the characteristic of the sampling units is in relation to the population mean. It is often convenient to express the magnitude of the population standard deviation in relation to the magnitude of the population mean. The ratio of the population standard deviation to the population mean is called the *coefficient of variation*. The coefficient of variation measures the *relative* population dispersion; the standard deviation measures the *absolute* population dispersion.

From the previous illustrations, it can be shown that the standard deviation of the population of recorded check amounts is \$54, and the coefficient of variation is, therefore, .4 (54/135). Contrasted to this, it can be shown that the standard deviation of the differences (audited amounts minus recorded amounts) is \$.68 and the coefficient of variation is 13.6 (.68/.05). This says the difference population has far less absolute dispersion (\$.68 vs. \$54), but is relatively more dispersed (13.6 vs. .4).

Sampling Distribution

From the population of sampling units a probability sample of size n is selected, a specified characteristic is observed for each sample item, and, based on these observed values, a point estimate is calculated for the corresponding population characteristic. The value of the estimates depends upon which sampling units are included in the sample. The *sampling distribution* describes the probability of each possible value of the sample point estimate. Knowledge of the sampling distribution allows the auditor to predict what will occur and evaluate what has occurred when sample estimates are made. Knowing the sampling distribution, at least approximately, allows the auditor to plan the extent of his observations and make useful inferences based on the observed results.

---

13. One reason for introducing symbols to represent population totals and means is that in many applications the auditor does not know the value of these quantities. For example, the *total audited amount* of a population (represented by X) is never known and likewise the total difference amount of a population (represented by D) is never known.

The concept of a sampling distribution may be illustrated by a very simple example. Assume the auditor is to select two accounts from the following five accounts and determine whether the balances are correct.

	Propriety of Balance	
Account	Correct	Incorrect
A	X	
B	X	
C		X
D	X	
E		X

Using unrestricted random sampling without replacement, the possible sample outcomes are as follows.

Accounts Selected	Incorrect Balances Selected
A, B	0
A, C	1
A, D	0
A, E	1
B, C	1
B, D	0
B, E	1
C, D	1
C, E	2
D, E	1

Since each of the combinations is equally likely to occur, the number of incorrect balances has a sampling distribution as follows.

Number of Incorrect Balances Observed	Number of Possible Outcomes	Probability of Outcome
0	3	.30
1	6	.60
2	1	.10
	10	1.00

In this simple illustration, the population occurrence rate is known to be .40 (two incorrect balances out of five balances). Even were this rate unknown, as is the usual case in practice, the auditor could make inferences about the population occurrence rate depending upon the sample outcome. Whenever no occurrences are observed in his sample of size two, for example, the auditor could infer that the population occurrence rate does not exceed .40. According to the sampling distribution there is a .70 probability of observing more than zero occurrences when the occurrence rate is .40. Consequently, the risk that this inference is incorrect never exceeds .30, regardless of what the population occurrence rate really is. The basis for this assertion is the fact that had the population occurrence rate exceeded .40, the probability of at least one observed error would have been at least .70 (.60 + .10 or 1.00 − .30).

The example shows that the sampling distribution depends upon (1) the sample size, (2) the method of selection, (3) the observed characteristic, (4) the evaluation procedure, and (5) the population distribution. Of these, the auditor selects (1) through (4), which permits him to control the probabilities that his inferences about (5) are incorrect.

As a more realistic example, if the auditor selects an unrestricted random sample of 50 purchase orders and determines whether each has been properly authorized, the number of orders with no authorization must be between 0 and 50. The sampling distribution specifies the probability of observing each of the 51 possible results. For each population occurrence rate, there is a unique sampling distribution describing the probability of observing each of the possible outcomes. For example, if the population of purchase orders contained no instances of orders lacking authorization, every sample that could be chosen would result in zero occurrences. The sampling distribution corresponding to a zero population occurrence rate would have a probability of 1.00 for zero sample occurrences. The following table shows the sampling distributions corresponding to population occurrence rates of .01 and .05.

Population Occurrence Rate			
.01		.05	
Number of Occurrences	Probability	Number of Occurrences	Probability
0	.6050	0	.0769
1	.3056	1	.2025
2	.0756	2	.2611
3	.0122	3	.2199
4	.0014	4	.1360
5	.0002	5	.0656
		6	.0260
		7	.0120

The probability column is interpreted as the frequency of observing the specified number of occurrences in repeated samples of size 50 from the population. For example, the frequency of observing zero occurrences in repeated samples of 50 is 60.5 percent when the population occurrence rate is 1 percent and only 7.69 percent when the population occurrence rate is 5 percent.

In auditing applications, the population occurrence rate is unknown—otherwise there would be no need to sample. Consequently, the auditor uses his knowledge of the possible sampling distributions to make an inference concerning the population occurrence rate based on the observed sample results. For example, if the auditor selects an unrestricted random sample of 50 purchase orders, and establishes that each one was properly authorized, it is relatively more likely that the observations came from a sampling distribution with a population occurrence rate of 1 percent than from a sampling distribution with a population occurrence rate of 5 percent.

Attributes

When the observed characteristic is an attribute, the population occurrence rate, sample size, and the method of selection determine the sampling distribution of the number of occurrences observed in the sample. Knowledge of the sampling distribution for the number of observed occurrences is sufficient for determining the sampling distribution of the usual estimator—the sample occurrence rate. The sample occurrence rate is equal to the number of occurrences divided by the sample size. If m represents the observed number of occurrences in a sample of n, the sample occurrence rate (p) is defined as the observed number of occurrences (m) divided by the sample size (n).

When the sample is selected using unrestricted random sampling without replacement, the appropriate sampling distribution is known as the *hypergeometric* distribution.[14] Extensive tables of this distribution do not exist, since separate tables would be required for each population size, sample size, and population occurrence rate. Consequently, several approximations are widely used. Among these, the binomial distribution is both well known and extensively tabled.[15] Although the binomial distribution is the exact sampling distribution only when sampling is with replacement, it nonetheless affords an excellent approximation of the hypergeometric distribution whenever the sample size is small relative to the population size. If the sample size exceeds, say, 10 percent of the population size, and sampling is without replacement, a correction factor will improve the approximation.

In some circumstances both the hypergeometric and binomial distributions can be approximated by a normal distribution.[16] For the binomial distribution this approximation is best when the population occurrence rate is equal to .50 and becomes less satisfactory the farther the occurrence rate is from .50. Since audit applications frequently involve population occurrence rates less than .10, the normal approximation is often inappropriate both in planning samples and in evaluating their results. The normal approximation can be considered satisfactory when either (1) the sample size is at least 60 and the population occurrence rate is between .30 and .70 or (2) the products of sample size times both the sample occurrence rate and its complement exceed 35.[17] This latter condition means that when the sample occurrence rate is about .10, the sample size should be at least 350. Because this sample size is very large for most audit applications, other approximations ordinarily should be used.

Another useful approximation is the Poisson distribution. This approximation is appropriate when both the population occurrence rate and the sampling fraction (sample size divided by population size) are small. The approximation is quite satisfactory when the population occurrence rate and the sampling fraction are both less than .10, and these conditions are common in auditing applications. When they exceed .10, correction factors can be used to improve the approximation. Relatively

14. A discussion of the hypergeometric distribution can be found in Dixon and Massey [8]. See chapter 9 for a computer time sharing program.
15. A discussion of the binomial distribution can be found in Dixon and Massey [8]. The following references contain tables of the binomial distribution: Aiken [1] and AICPA [2], vols. 2 and 6.
16. The normal distribution is discussed further later in this chapter.
17. These conditions are suggested in Hansen, Hurwitz, and Madow [12], p. 131.

compact but nonetheless comprehensive tables for the Poisson distribution are readily available.[18]

Variables

When the observed characteristic is a variable (for example, an audited dollar amount, a difference amount, or a ratio of audited amounts to recorded amounts), no single sampling distribution will apply to every situation. Rather, the shape of the applicable sampling distribution of a specified variable depends on (1) the shape of the population distribution, (2) the method of sample selection, and (3) the sample size.

To illustrate, suppose an unrestricted random sample were chosen from the population of 500 checks previously described and the recorded amount of each check in the sample observed. Suppose the population total (total recorded amount) is estimated by multiplying the population size (500) by the sample mean ($\bar{y}$).[19] When the sample size is 1, the sampling distribution is the same as the population distribution shown in figure 2.1, except that the limits of the class intervals are each multiplied by 500. Consequently, there is a .35 probability that the estimate is between \$500 and \$25,000, a .15 probability that it will be from \$25,500 to \$50,000, and so forth.

As the sample size is increased, the sampling distribution changes shape, becoming more symmetrical and bell-shaped. This is illustrated by the following graph based on selecting many unrestricted random samples of size 50 from the population. To make the scale comparable to figure 2.1, the frequency of the sample means is graphed, rather than 500 times the sample means.

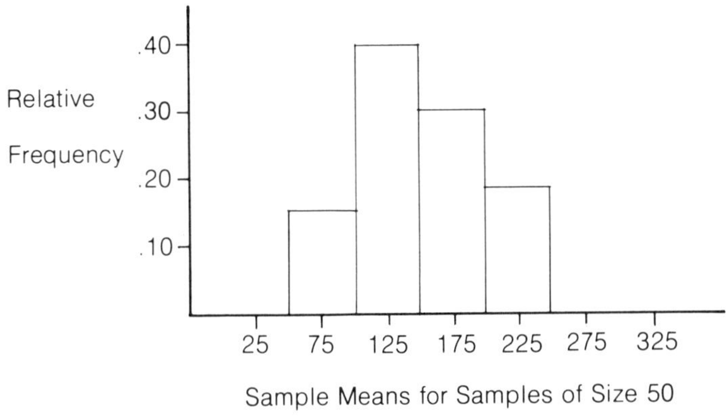

Figure 2.4

The sampling distribution shown in figure 2.4 is much more symmetrical than the corresponding population distribution shown in figure 2.1, which is highly skewed. Sample size is responsible for this phenomenon. For any population, the sampling distribution of the sample means is nearly symmetrical when the sample is large enough.

The mean value of the sampling distribution in figure 2.4 is \$135. This is also the

18. See Molina [16].
19. The sample mean, $\bar{y}$, is defined in a manner similar to the population mean. If $y_1, y_2, y_3, \ldots, y_n$ denote the n sample values,

$$\bar{y} = \frac{\sum y_i}{n}.$$

value of the population mean. The mean of the sampling distribution always equals the population characteristic whenever the estimator is unbiased. The term *unbiased* means that the average of the estimates computed for all possible samples equals the corresponding population characteristic.

The standard deviation of the sampling distribution is called the *standard error of the estimate.* The standard error in figure 2.4 is about $7.25. The dispersion of the sampling distribution is much less than that of the population (population standard deviation equals $54). The amount of reduction is determined by the sample size. As a general rule, the standard error of the estimate equals the population standard deviation divided by the square root of the sample size and multiplied by the so-called finite population correction factor. This factor, which equals the square root of 1.0 minus the sampling fraction (sample size divided by population size), is only important when the sample size exceeds 10 percent of the population.

For example, the standard error in figure 2.4 was computed as

$$\$7.25 = \frac{\$54}{\sqrt{50}}\sqrt{1 - \frac{50}{500}}.$$

Neglecting the finite population correction factor, the standard error would be computed as

$$\$7.64 = \frac{\$54}{\sqrt{50}}.$$

The reduction of 5 percent corresponds to a sampling fraction of 10 percent. With a smaller sampling fraction, such as 3 percent, the reduction is about 1.5 percent. In general, use of the finite population correction factor will reduce the standard error by about one-half the sampling fraction.

The auditor can control the magnitude of the standard error by selecting an appropriate sample size. This is a key element in planning. Neglecting the finite population correction factor, the standard error goes down by the square root of the sample size. For example, to reduce the standard error by a factor of two, the sample size needs to be increased fourfold.

Normal Distribution

The shape of the sampling distribution will resemble the normal distribution when the sample size is sufficiently large. This is a general result applicable to each of the techniques discussed in chapters 5 and 6. More precisely, statistical theory says that the distribution of the standardized estimator (the estimate minus its mean and divided by the standard error) is approximated by a standardized normal distribution with a mean of zero and a standard deviation of one. This is a fundamental result that underlies most of the precision and reliability statements associated with variable estimators. Because they affect the statistical validity of the auditor's results, it is important that sample sizes be large enough to warrant the normal approximation. This important topic is addressed for each estimator in chapters 5 and 6.

A related issue arises from the fact that the standard error of the estimate must be estimated from the observed sample results. Consequently, the sample must be large enough to obtain a *stable* estimate of the standard error as well as to warrant the approximate normality of the standardized estimator when the estimated standard error is substituted for the true but unknown standard error. At times, the auditor can use a distribution that is different from, but related to, the normal—this is the so-called *Student's t-distribution.*[20]

The normal distribution plays a central role in the statistical evaluation of estimates of the total audited amount whether based on the sample mean, difference, ratio, or regression. A normal distribution is based on a mathematical formula that depends solely on the specified mean and standard deviation. Figure 2.5 shows a normal distribution corresponding to a mean of 100 and a standard deviation of 10. Probabilities are measured by the *area* under the curve (the total area is, of course, equal to 1.0).

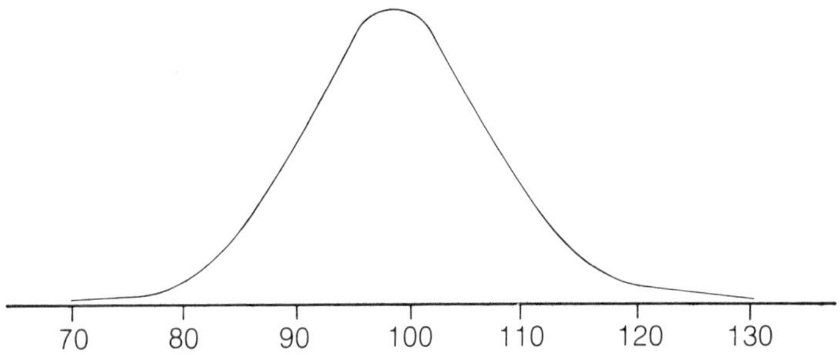

Figure 2.5

In this example, the probability of observing a value between 90 and 110 is nearly .68. The lower limit (90) represents one standard deviation (10) below the mean (100) and the upper limit (110) represents one standard deviation above the mean. The probability of being within two standard deviations of the mean (80 to 120 in this example) is about .95 (.9546 to be exact). The probability of being within three standard deviations of the mean is nearly certain—being equal to .9972.

These probabilities will always be the same for any normal distribution and are easily determined from a normal table (see Appendix 3). The entries in the body of the table represent the area (probability) from the *mean* to a specified number of standard deviations. For example, the entry corresponding to 1.64 standard deviations is .4495. This says that the probability of observing a value between the mean and 1.64 standard deviations above the mean is .4495. Since the normal curve is symmetric about the mean, .4495 also represents the probability of observing a value between the mean and 1.64 standard deviations below the mean.

Multiplying the entries by two (2) gives the probability of being within the specified number of standard deviations of this mean. For example .8990 (.4495 × 2) represents the probability of observing a value within 1.64 standard deviations of the mean (from 1.64 standard deviations below the mean to 1.64 standard deviations above the mean).

Since the area on either side of the mean of a normal distribution is equal to one-half (.5000), adding .5000 to the entries in the table yields the probability of observing a value *not greater* than the specified number of standard deviations *above* the mean or *greater* than the specified number of standard deviations *below* the mean. For

20. See Appendix 3 for a table showing this distribution.

example, .9495 represents the probability of observing a value not greater than 1.64 standard deviations *above* the mean (or greater than 1.64 standard deviations below the mean).

The auditor uses these normal probabilities to determine his audit risks. A detailed explanation of this is contained in the following chapter.

3

Precision, Reliability, and Sampling Risk

Decisions based on samples involve risk. The auditor who uses statistical sampling to estimate a population characteristic needs to know how good his estimate is. The sampling distribution provides the basis for calculating the magnitude of risk as well as measuring how accurate the estimate is. The key concepts used in the calculations are the *precision* and *reliability* of an estimate. This chapter discusses these concepts and applies them to measuring the sampling aspect of audit risk.

Precision and Reliability

Section 320A.03 of SAS no. 1 states:

> Statistical samples are evaluated in terms of "precision," which is expressed as a range of values, plus and minus, around the sample result and "reliability" (or confidence), which is expressed as the proportion of such ranges from all possible similar samples of the same size that would include the actual population value.

Precision is a measure of the closeness between the sample estimate and the corresponding unknown population characteristic. Reliability measures the frequency with which the difference between the sample estimate and the population value does not exceed the precision. For example, a reported precision of $50,000 at a reliability of .97 indicates that, among all possible samples, 97 percent will have a difference between the sample estimate and the population characteristic less than or equal to $50,000.

The precision of any estimate at a specified reliability is determined by the sampling distribution of the estimate. For example, because the sampling distribution of an estimate of the population audited amount is approximately normal when the sample size is sufficiently large, properties of the normal distribution described in the previous chapter may be used by the auditor to calculate the precision of his estimate at any specified reliability.

When the estimate is unbiased, the mean of the sampling distribution is the unknown population audited amount. This implies that the *difference* between the estimate and the population audited amount is approximately normal, but with a mean of *zero;* that is, regardless of the value of the unknown population audited amount, the sampling distribution of this difference is centered at zero.

The standard deviation of this sampling distribution is the standard error of the estimated audited amount. This standard error can be estimated from the sample observations. For example, if the estimated standard error is $20,000, then the sampling distribution of the difference between the estimate and the total audited amounts is distributed approximately as the normal distribution shown in figure 3.1.

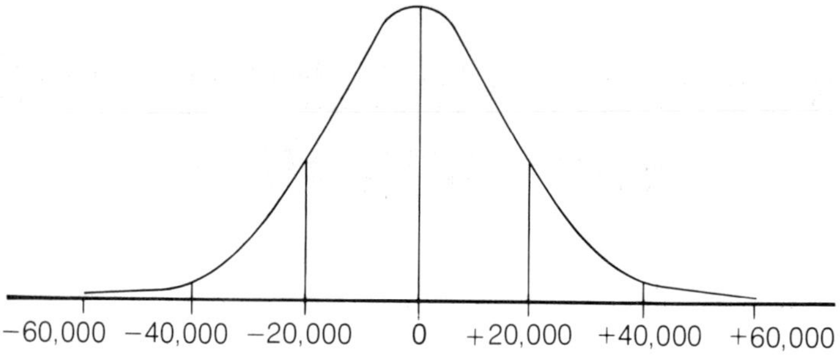

Figure 3.1

The normal tables show that the probability that the difference is less than or equal to $20,000 is .6826 (equal to the area under a normal curve from one standard deviation below the mean to one standard deviation above the mean). Another way this may be stated is that a reliability of .68 is associated with a precision of $20,000 (the estimated standard error).

Specifying a reliability of .90, the auditor needs to determine the number of standard deviations from the mean required to contain an area of .90. From the normal table, this is determined to be 1.65. Continuing with this example, the precision is equal to $33,000 (1.65 × $20,000) at a reliability of .90. Similarly, the precision is $39,200 (1.96 × $20,000) at a reliability of .95. In all these cases, the precision is obtained by multiplying the estimated standard error ($20,000) by the appropriate number of standard deviations corresponding to a specified probability. Auditors often use a condensed table of such factors, labeled U_R, corresponding to a specified reliability, labeled R. Such a table appears in Appendix 3.

Determining the probability that the difference between the estimate and the corresponding population characteristic is less than a specified amount can be done as above without regard to the algebraic sign of the difference or as a *signed* difference. Defining a signed difference involves specifying how the difference is to be calculated; the convention here is to make the difference equal to the population characteristic minus the estimate. Using this convention, a *positive* difference corresponds to the estimate's being *less* than the population characteristic while a *negative* difference means the estimate *exceeds* the population characteristic.

Using the example shown in figure 3.1, the auditor can compute the probability that the signed difference is not more than $20,000. This probability corresponds to the area under the normal curve to the left of $20,000, which is equal to .8413. This may be interpreted as saying that the signed difference has a precision of $20,000 at a reliability of .8413. To avoid confusion, this is called the *upper* precision at a *one-sided* reliability. The term *upper* refers to the fact that, in this case, $20,000 is an upper limit to the signed difference; the term *one-sided* is used as a reminder that the relevant area is all the area under the curve to the left of $20,000.

An upper precision of $33,000 corresponds to a one-sided reliability of .95, while an upper precision of $39,200 corresponds to a one-sided reliability of .975. This illustrates the fact that any stated precision (for example, $33,000) at a specified interval reliability R (.95) also may be interpreted as the same upper precision at a different one-sided reliability which is labeled R_1. The relationship between the two is expressed by the following formula:

$$R_1 = \frac{R + 1.0}{2}.$$

For example, an interval reliability of .95 corresponds to a one-sided reliability of .975 (1.95/2).

It is also possible to calculate a *lower* precision for an estimate. For example, the probability is .8413 that the signed difference exceeds −$20,000. The negative sign is required because the calculation involves the signed difference. In this case, the lower precision is $20,000 at a one-sided reliability of .8413.

The preceding illustration has focused attention on the difference, signed or not, between the sample estimate and the corresponding population characteristic. In practice, the auditor may require a *precision interval* or an *upper (lower) precision limit* that involves the sample estimate and its precision.

A *precision interval* is formed by adding and subtracting the precision at a specified interval reliability to the sample estimate.[1] For example, if the sample estimate of the total audited amount is $650,000 with a precision of $33,000 at an interval reliability of .90, then the precision interval ranges from $617,000 ($650,000 − $33,000) to $683,000 ($650,000 + $33,000). A precision interval contains the population total audited amount with a specified reliability.

Proper interpretation of this result is very important in practice. What does it mean to assert that a precision interval contains the population characteristic with a specified reliability? It means that were the auditor to compute a precision interval for each possible sample, the proportion of intervals containing the population characteristic would equal the specified reliability. When the auditor asserts that the reliability of his precision interval is .95, he is stating that the statistical *procedure* he used results in the population characteristic's being contained within the precision interval 95 times out of 100.

An upper precision limit may be computed by adding the upper precision to the sample estimate. If, as in the previous example, the sample estimate of the total audited amount is $650,000 with a precision of $33,000 at .90 interval reliability, then the upper precision is $33,000 at .95 one-sided reliability and the upper precision limit is $683,000 ($650,000 + $33,000) at the one-sided reliability of .95. The upper precision limit exceeds the population characteristic with a specified one-sided reliability.

In a similar manner, the auditor may calculate a lower precision limit by subtracting the *lower* precision from the sample estimate. The lower precision limit in the example would be $617,000 at a one-sided reliability of .95 because the lower precision is $33,000 at a one-sided reliability of .95, and $650,000 − $33,000 = $617,000.

1. A *precision interval* is also known as a *confidence interval* and the associated *reliability* is also called the *confidence level*.

The validity of these illustrated calculations depends upon two properties of the sampling plan:

1. The approximate normality of the sampling distribution.
2. Availability of a stable estimate of the standard error from the sample.

In practice, failure of either of these might result in meaningless calculations. The survey in chapters 5 and 6 of statistical techniques that depend upon these two properties includes a discussion of conditions that assist the auditor in recognizing questionable applications.

Attributes

Some statistical techniques, such as attribute methods, do not depend upon approximate normality of the sampling distribution. In such cases, the concepts of precision and reliability remain the same, but their method of calculation changes. When the observed characteristic is an attribute, the sampling distribution of the observed number of occurrences in an unrestricted random sample of size n may be approximated by the binomial distribution. This distribution depends solely upon the unknown population occurrence rate. Because the distribution is not symmetric, upper precision will not equal lower precision, and consequently the technique of adding the precision to or subtracting it from the estimate does not apply as it did with the normal distribution.

To illustrate the determination of an upper precision limit, assume the auditor has observed one occurrence in an unrestricted random sample of 80 and specifies a one-sided reliability of .99. The auditor determines from the binomial tables the probability of observing more than one occurrence in a sample of 80 for several population occurrence rates. For example, the probability is .696 when the occurrence rate is .03, the probability is .914 when the occurrence rate is .05, and the probability is .990 when the occurrence rate is .08. From the set of population occurrence rates the auditor selects the one that equals or exceeds the specified one-sided reliability of .99. In this example, the selection would be .08 as the upper precision limit.

What is the reasoning behind this choice? If the population occurrence rate were as large as .08, the probability that the auditor would have observed more than one occurrence is .99. Consequently, either an event of very small probability has occurred (only one occurrence in a sample of 80) or the population occurrence rate does not exceed .08. Selecting the latter alternative explanation, the auditor has a one-sided reliability of .99 of being correct.

Determining an upper precision limit for any observed number of occurrences in an unrestricted random sample is simply a matter of finding the population occurrence rate that makes the probability of observing more than the observed number of occurrences equal to the desired one-sided reliability. The upper precision limit depends upon (1) the sample size, (2) the observed number of occurrences, and (3) the specified one-sided reliability.

A lower precision limit can also be determined for any observed number of occurrences. The lower precision limit corresponding to any observed number of occurrences is the population occurrence rate that makes the probability of observing fewer than the observed number of occurrences equal to the desired one-sided reliability. This procedure must be modified when the auditor observes zero occurrences; in this case, the lower precision limit is set equal to zero.

The upper precision of an attribute estimate is equal to the upper precision limit minus the sample occurrence rate. The lower precision is equal to the sample occurrence rate minus the lower precision limit. For example, one occurrence in a sample of 80 yields an occurrence rate of .0125 and an upper precision limit of .08 at a one-sided reliability of .99. Therefore, the upper precision equals about .07 (.08 − .0125). The lower precision limit is nearly zero and, consequently, the lower precision is about .01 (.0125 − 0).

The precision interval ranges from the lower precision limit to the upper precision limit. The reliability of the precision interval is related to the one-sided reliability, as previously described. That formula may be rewritten in the following way to express the interval reliability R in terms of the one-sided reliability R_1:

$$R = 2R_1 - 1.0.$$

Thus, when one-sided reliability (R_1) is .99, as in the example, the reliability of the precision interval (R) is .98.

The meaning of precision and reliability remains the same for attributes as it is for variables. For example, a one-sided reliability of .99 means that 99% of all possible samples of a particular size will yield an upper precision limit that exceeds the actual population occurrence rate. Similarly, when the interval reliability of a precision interval for a particular size sample is .90, then 90 percent of all samples of that size yield precision intervals that contain the population occurrence rate.

Statistical Objective

When the statistical evidence gives an estimate of a population characteristic together with the (estimated) precision of that estimate at a specified interval reliability level (R), the auditor can use the results in either of two ways—for estimation or decision-making.

Estimation Objective

The auditor can use an estimation objective whenever he needs to construct a value that estimates the unknown population characteristic. For example, the auditor might want to estimate the fraction of a finished-goods inventory that has not been sold within the last six months; estimate the amount of overstatement in the inventory in order to propose an adjustment to the recorded amount; estimate an allowance for bad debts. These and other examples have the common property of being constructive; that is, an estimation objective is used to construct an estimate of some characteristic of the accounting system.

To estimate a population characteristic, the auditor needs to specify both a desired precision and a desired interval reliability level. The sample can then be designed and the sample size determined to reflect those selected specifications. When the characteristic is an attribute, the auditor needs to specify an anticipated occurrence rate in order to determine the appropriate sample size; this contrasts to variables sampling, in which an additional ingredient is the estimated standard deviation of the characteristic of interest. For example, when the observed characteristic is the monetary difference between audited and recorded amounts, the sample size required to achieve a desired precision depends upon the estimated standard deviation of differences among the sampling units in the population.

For an estimation objective, the precision measures the maximum probable difference between the estimate and the corresponding population characteristic. At a given interval reliability level (*R*), the smaller the precision, the closer these two amounts will be. Thus, the reliability level is an essential ingredient in defining the maximum probable difference.

The achieved precision depends upon the observed sample results. Consequently, the achieved precision will be the same as the planned precision only when the anticipated occurrence rate equals the observed occurrence rate in attributes estimation or when the sample standard deviation of the observed characteristic equals the estimated standard deviation used in planning a variables estimation sample. This will not ordinarily occur unless the auditor has advance information concerning the population distribution of the observed characteristic. In most applications, the auditor needs to use the best information available to obtain a conservative estimate of the relevant anticipated occurrence rate or estimated standard deviation, so that the achieved precision will be no larger than planned.

Decision Objective

Much of the auditor's work is not constructive, but critical.[2] He must decide whether the evidence supports such propositions as compliance with the pertinent accounting control is satisfactory, this inventory amount is not materially misstated, or these accounts receivable represent bona fide amounts owed to the client. In these circumstances the auditor must decide whether or not the statistical evidence supports the proposition. The term *decision-making* is used to describe this use of a statistical sample.[3]

To decide whether a statistical sample supports a stated proposition, the auditor must specify a decision rule. This rule must be constructed so that no matter what sample results are observed, the statistical sample either does or does not support the proposition. In formulating the rule, the auditor controls the risks of making a mistake. Two kinds of mistakes are possible: (1) deciding that the evidence supports the proposition when, in fact, the proposition is not true, and (2) deciding that the evidence fails to support a proposition when, in fact, the proposition is true. When the decision is based on any kind of sample, the possibility of deciding incorrectly can never be eliminated, but the auditor using statistical sampling can control the risks of error by appropriately selecting the precision and reliability of the estimate.

Chapter 4 discusses control of these risks when attributes methods are used. The following discussion concerns control of the risks when the variables sampling methods discussed in chapters 5 and 6 are used, as they frequently are, for decision-making in connection with substantive audit tests.

Positive Approach. In applying statistical decision-making to substantive tests, the auditor must select one of two possible propositions. The first, called the *positive approach,* hypothesizes that the recorded amount is correct. The decision rule for the

---

2. See Mautz [15].
3. In statistical theory this is known as *hypothesis testing,* a term employed by Elliott and Rogers [9]. Since testing is already used extensively in auditing, the term *decision-making* was adopted as descriptive of how the sample evidence is used.

positive approach specifies under what circumstances the statistical evidence supports the correctness of the recorded amount. The decision rule is described as follows.

Compute the value of the estimated audited amount, together with its achieved precision at a specified interval reliability. If the recorded amount is within the calculated precision interval, decide that the statistical evidence supports the correctness of the recorded amount. If not, decide that the evidence fails to support the correctness of the recorded amount.

The risks of deciding incorrectly are controlled by selecting the values of the precision and the reliability. Specifically, the risk that the evidence erroneously fails to support the correctness of the recorded amount when it is in fact correct, is equal to the complement of the reliability (that is, $1.0 - R$, where R represents the reliability). In statistical terminology this is called the *alpha risk* (α).

The more important error is erroneously deciding that the statistical evidence supports the correctness of the recorded amount when in fact the recorded amount is not correct. Of course, a recorded amount can be incorrect by varying amounts ranging from small to large. In exercising control over this type of error, the auditor must specify the amount of misstatement that is important to him. This is closely related to the amount of misstatement he would regard as *material;* indeed, the term "material amount of misstatement" will be used to describe this amount.

The probability that the statistical evidence erroneously supports the correctness of the recorded amount when the recorded amount is misstated by a material amount is called the *beta risk* (β). This risk is controlled by selecting the magnitude of the precision in relation to the material amount of misstatement. When the interval reliability is .95 (alpha risk is .05), for example, setting precision equal to one-half the material amount results in a beta risk of .025, while setting precision equal to the material amount results in a beta risk of .50. To determine the desired size of the precision the auditor can use either a table (see Appendix 3D) or a formula.

The formula may be expressed as follows:

$$A = \frac{z_{\alpha/2} M}{z_{\alpha/2} + z_{\beta}}.$$

A is the desired precision corresponding to an interval reliability (R), the interval reliability (R) equals the complement of the alpha risk ($R = 1 - \alpha$), M is the minimum amount considered material, $z_{\alpha/2} = U_R$ and z_{β} = the normal table value of $(.5000 - \beta)$.

Alternatively, the formula may be expressed in the following manner by replacing $z_{\alpha/2}$ by U_R:

$$A = \frac{U_R M}{U_R + z_{\beta}}.$$

As an example, suppose an auditor is testing a raw material inventory with respect to quantities, prices, and extensions. The recorded amount is $1 million. He decides that a material amount would be $100,000—this is based on the financial statements taken as a whole. Moreover, he determines that a beta risk of .10 and an alpha risk of .05 would be appropriate in the circumstances.

Using the formula, he finds that

$$A = \frac{1.96 \times \$100{,}000}{1.96 + 1.28}$$

$$= \$60{,}494 \text{ or } \$60{,}500 \text{ rounded.}$$

In this case, 1.96 is the normal table value of $z_{.025}$ ($z_{.05/2}$) and 1.28 is the normal table value closest to .4000 (.5000 − .1000). The decision rule for this example follows.

Compute the value of the estimated audited amount. If this amount is within \$60,500 of the recorded amount, \$1 million, decide that the statistical evidence supports the correctness of the \$1 million with respect to quantities, prices, and extensions.

This rule can also be stated in terms of the point estimate being between \$939,500 (\$1,000,000 − \$60,500) and \$1,060,500 (\$1,000,000 + \$60,500). Such an equivalent restatement is always possible. The resulting interval is from the recorded amount minus the precision to the recorded amount plus the precision. This interval is called the planned *decision interval.*

Figure 3.2 illustrates the alpha risk in this example. The illustrated sampling distribution is approximately normal. When the recorded amount is correct, the mean of the sampling distribution equals the recorded amount (\$1 million). The alpha risk is the probability that the estimate differs from the recorded amount (\$1 million) by more than the precision (\$60,500). This occurs when the estimate lies outside the decision interval between \$939,500 and \$1,060,500. Because the recorded amount is assumed to be correct, the probability is just the shaded areas shown which added together equal .05.

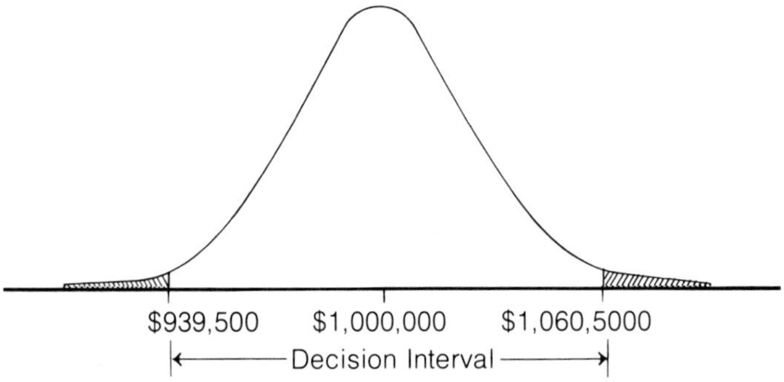

Figure 3.2

The beta risk is similarly illustrated by figure 3.3. For this example, the beta risk is the probability that the estimate is within \$60,500 of the recorded amount (\$1 million) when the true amount is either overstated or understated by a material amount (\$100,000). When the recorded amount is overstated by \$100,000, the true amount of \$900,000 equals the mean of the sampling distribution. The beta risk of .10 is shown by the shaded area.

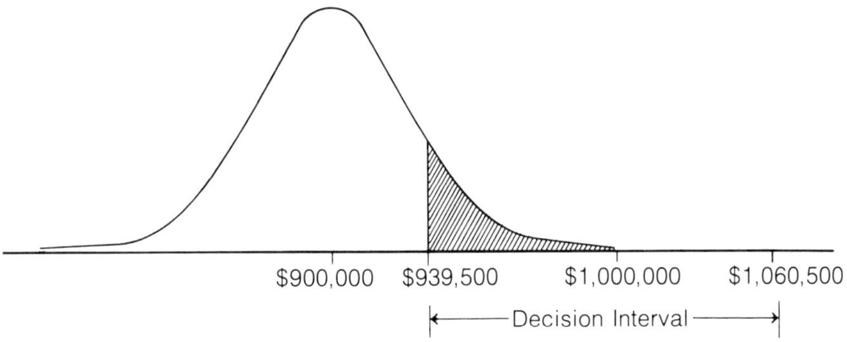

Figure 3.3

Figure 3.4 illustrates the same beta risk when the recorded amount is understated by $100,000, and, therefore, the mean of the sampling distribution equals $1,100,000 ($1,000,000 + $100,000).

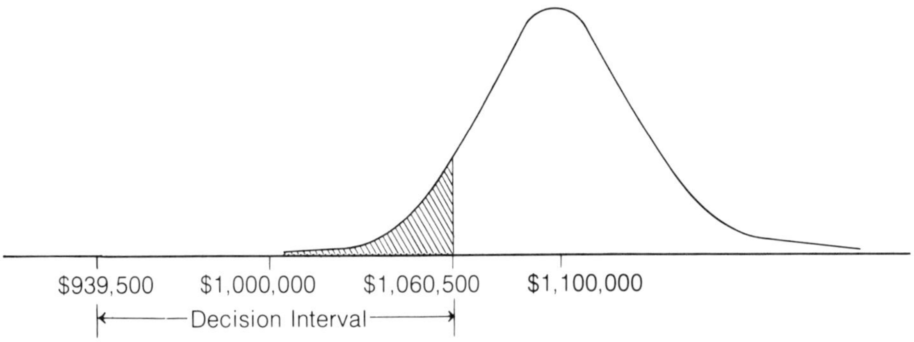

Figure 3.4

Whether the rule is stated in terms of the precision interval's containing the recorded amount or the estimated audited amount's being in the decision interval, it is necessary that the rule be strictly followed. Otherwise the actual risks may differ from the planned risks, and the auditor would lose one of the chief benefits of using statistical sampling.

In addition to adhering to the decision rule, the auditor can adjust the precision whenever the achieved precision differs from the planned precision. This occurs when the achieved standard error differs from the standard error used in planning. Without an adjustment, the effective beta risk differs from the planned beta risk whenever the achieved precision is not equal to the planned precision. If the achieved precision is smaller than planned, the effective beta risk is smaller, while an achieved precision larger than planned results in a higher effective beta risk. The alpha risk would not be affected by any difference between planned and achieved precision.

To maintain the effective beta risk at the planned level, the auditor can calculate an adjusted precision based upon the planned precision, the achieved precision, and the material amount of misstatement. The adjusted precision equals the achieved precision plus the material amount multiplied by the difference between the planned precision and the achieved precision divided by the planned precision. In symbols this may be expressed as follows:

$$A'' = A' + M\left(\frac{A - A'}{A}\right),$$

where M represents the material amount, A represents the planned precision, A' represents the achieved precision, and A'' represents the adjusted precision.[4]

For example, if the planned beta risk is .20, the planned alpha risk is .05, and the planned precision is \$700,000 based on a material amount of \$1 million, the effective beta risk is .025 when the achieved precision is \$500,000 at the planned reliability of .95. The adjusted precision that maintains the effective beta risk at .20 is equal to

$$\$785{,}714 = \$500{,}000 + \$1{,}000{,}000\left(\frac{\$700{,}000 - \$500{,}000}{\$700{,}000}\right).$$

A consequence of using the adjusted precision to maintain the effective beta risk at the planned level is that the effective alpha risk changes from the planned level. When the adjusted precision is larger than the planned precision, the effective alpha risk decreases (the reliability increases). Conversely, when the adjusted precision is smaller than the planned precision, the effective alpha risk increases. In the illustration, the adjusted precision of \$785,714 exceeded the planned precision of \$700,000, and, consequently, the alpha risk is lower than planned. To compute the effective alpha risk, an adjusted reliability factor is calculated. The adjusted reliability factor equals the planned reliability factor multiplied by the ratio of the adjusted precision to the achieved precision. In the example the planned reliability factor equals 1.96; consequently, the adjusted reliability factor equals 3.08 ($1.96 \times (785{,}714/500{,}000)$). This corresponds to a reliability in excess of .99 and an effective alpha risk of less than .01.

The same procedure may also be used to make an adjustment when the achieved precision is larger than the planned precision. For example, suppose the auditor planned an alpha risk of .05 and a beta risk of .025 and also specified a desired precision of \$500,000 where \$1 million represented a material amount. Now, an achieved precision of \$700,000 would lead to an adjusted precision of \$300,000 (\$700,000 + \$1,000,000[(\$500,000 − \$700,000)/\$500,000]). The reliability associated with this adjusted precision is smaller and its complement, the alpha risk, is therefore larger. The adjusted reliability factor is .84 ($1.96 \times 300{,}000/700{,}000$) for which the effective reliability is .60 and the effective alpha risk is .40 (1.0 − .60).

Allowing the effective alpha risk to increase, the auditor incurs an increased probability that the sample evidence will erroneously fail to support the correctness of the recorded amount when, in fact, the recorded amount is correct. When the estimate falls outside the adjusted decision interval, the auditor must obtain additional evidence; one possible source of such evidence would be additional sample ob-

4. This formula applies when both the planned and achieved precisions use the same reliability factor. When the achieved precision uses a factor from the Student's t-table, a somewhat different formula is applicable. See Appendix 5A. That appendix also outlines a procedure for testing the possibility that the sample estimate of the standard error is too low. Also, it is possible to avoid this calculation by stating the decision rule as deciding the statistical evidence supports the recorded amount whenever the absolute difference between the estimated audited amount and the recorded amount does not exceed the material amount (M) minus z_β times the achieved standard error of the estimate.

servations. Obtaining additional evidence involves increased costs and time, but that is a better alternative than allowing the beta risk to increase beyond its planned level.

Negative Approach. In following the *positive* approach the auditor starts with the proposition that the recorded amount is correct and uses the statistical evidence to support or reject that proposition. In contrast to this is the *negative* approach by which the auditor begins with the proposition that the recorded amount is incorrect by a material amount and uses the statistical evidence to support or reject that proposition. While it can be demonstrated that the two approaches are equivalent, the negative approach is assumed throughout section 320B of SAS no. 1.

Following the negative approach, the auditor formulates a proposition that he believes is not correct—the existence of a material understatement or overstatement in the recorded amount. When the statistical evidence renders the proposition implausible, he decides that it is not true.[5] That is, when the statistical evidence fails to support the potential existence of a material amount of error, the auditor decides that the recorded amount is not materially incorrect. A decision rule following this approach is the following.

Compute an estimate of the total difference between the audited amount minus the recorded amount together with the precision of the estimate at a specified interval reliability. If the upper precision limit (the estimated difference plus the precision) is smaller than the material amount and the lower precision limit (the estimated difference minus the precision) is larger than the negative of the material amount, decide that no material misstatement exists. Otherwise, decide that the recorded amount may be materially misstated.

This decision rule means that the evidence supports the position that no material misstatement exists whenever the estimated difference in absolute terms is less than the material amount (M) minus the precision. This rule also says that the decision that no material misstatement exists is supported by the statistical evidence whenever the estimated difference regardless of sign, plus the precision is less than a material amount.

When the auditor uses interval reliability (R), the probability that this negative approach results in erroneously deciding that no material misstatement exists, when, in fact, the amount of overstatement or understatement is equal to a material amount, is equal to the complement of the interval reliability divided by 2.0 $[(1.0 - R)/2]$. Thus, when the auditor uses an interval reliability (R), the beta risk (β) is represented as[6]

$$\beta = \frac{1.0 - R}{2}, \text{ or, solving for } R,$$

$$R = 1.0 - 2\beta.$$

For example, if the auditor sets the tolerable beta risk at .15, then $R = .70$ $(.70 = 1.0 - 2 \times .15)$.

---

5. This is the procedure described in statistical theory under testing statistical hypotheses as *rejecting the null hypothesis*.
6. Throughout this book, *beta risk* will always refer to the risk that the auditor decides that no material monetary error exists when, in fact, the recorded amount is materially misstated. This is contrary to standard statistical practice when the negative approach is used.

Instead of using an interval reliability (R), the auditor may apply the above decision rule using a one-sided reliability (R_1). When this is done, the beta risk (β) is simply represented as the complement of the one-sided reliability (R_1). In symbols,

$$\beta = 1.0 - R_1.$$

For example, a tolerable beta risk of .15 would entail using a one-sided reliability of .85.

Section 320B.30 of SAS no. 1 states:

> The risk that material errors will not be detected in the auditor's examination is measured by the complement of the reliability level used if the auditor compares the upper precision limit of monetary error to the amount he considers material. This is the basis for the discussion pertaining to reliability in subsequent paragraphs. On the other hand, if the auditor adopts the decision rule to accept the book value as materially correct only if it is included in the statistical precision range, this constitutes a hypothesis test and he should interpret the following paragraphs in that context.

The first two sentences quoted above describe what is called here the negative approach, using one-sided reliability; on the other hand, the last sentence refers to the positive approach. Because all references to reliability in the subsequent paragraphs of section 320B assume one-sided reliability (R_1) associated with the negative approach, those references should be interpreted as meaning the complement of the beta risk when the positive approach is used.

The second type of risk is erroneously deciding that there may be material misstatement when the recorded amount is, in fact, correct. This is called the alpha risk (α). Using the negative approach, this risk can be controlled by selecting the precision as a fraction of the material amount. This can be done either by using a table (Appendix 3D) or by using the following formula:

$$A = \frac{z_\beta}{z_{\alpha/2} + z_\beta} M,$$

when A is the desired precision, M is the minimum amount considered material, $z_{\alpha/2}$ is the normal table value of $(.5 - \alpha/2)$, and z_β is the normal table value of $(.5 - \beta)$. When the auditor uses a one-sided reliability (R_1), $z_\beta = U_{R_1}$ and so the formula can be rewritten

$$A = \frac{U_{R_1}}{z_{\alpha/2} + U_{R_1}} M.$$

For example, suppose the negative approach is applied to the previous inventory example with a recorded amount of \$1 million and a material amount of \$100,000. The auditor specifies a beta risk of .10 (the risk of erroneously deciding there is no material monetary error when the overstatement is \$100,000 or the understatement is \$100,000). This means that the auditor can use a one-sided reliability of .90. He also specifies .05 as the alpha risk (erroneously deciding there may be material misstatement when the recorded amount is correct). Using the formula, $z_\beta = U_{R_2} = 1.28$, $z_{\alpha/2} = 1.96$, and $M = \$100{,}000$. Therefore,

$$A = \frac{1.28}{1.96 + 1.28} \$100{,}000$$

$$= \$39{,}506 \text{ or } \$39{,}500 \text{ rounded.}$$

Applying the decision rule for the negative approach the auditor will decide that a material error may exist if the upper precision limit (estimated total difference plus $39,500) exceeds $100,000 or the lower precision limit (estimated total difference minus $39,500) is less than −$100,000. These conditions are equivalent to the estimated total difference's exceeding $60,500 ($100,000 − $39,500) or being less than −$60,500.

Figure 3.5 illustrates the risk that the auditor decides there may be material error when the recorded amount is correct. In this circumstance the sampling distribution of the total difference is centered at zero and the shaded areas indicate the resulting probability that the estimated total difference either exceeds $60,500 or is less than −$60,500.

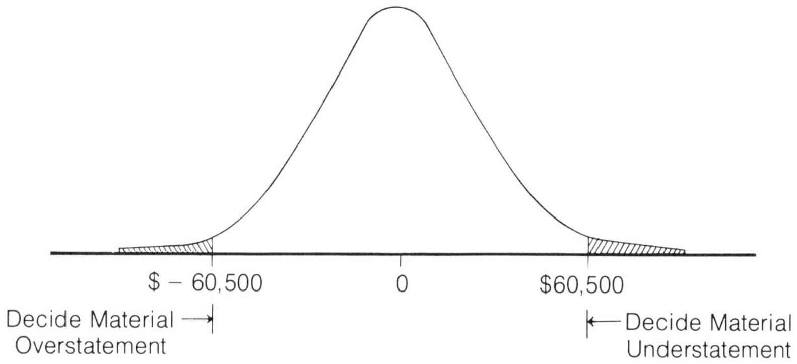

Figure 3.5

Figure 3.6 illustrates the risk that the auditor may decide there is no material error when, in fact, the amount of overstatement is $100,000. Because differences are computed as the audited amount minus the recorded amount, a negative sign indicates an overstatement. When the sampling distribution is centered at −$100,000, the probability that the estimated total difference is greater than −$60,500 and less than +$60,500 is indicated by the shaded portion under the curve.

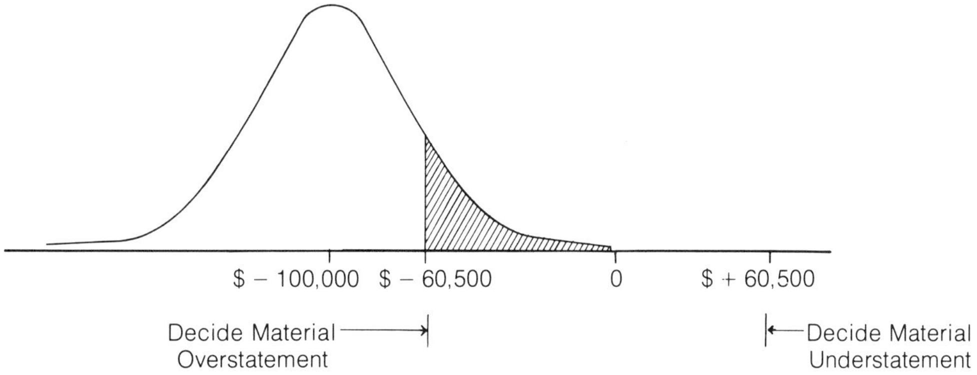

Figure 3.6

Figure 3.7 illustrates the same risk when the recorded amount is understated by $100,000.

When the achieved precision does not equal the planned precision under the negative approach, the auditor does not need to make any adjustment to maintain the effective beta risk of erroneously deciding that there is no material misstatement

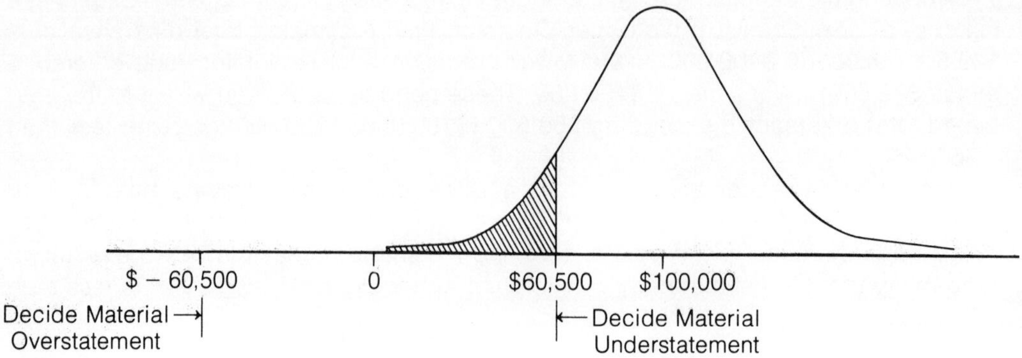

Figure 3.7

at the planned level. Of course, the effective alpha risk of erroneously deciding that there may be a material misstatement is changed from the planned level when the achieved precision differs from the planned precision. If the achieved precision is larger than planned, the risk is greater; while, if the achieved precision is smaller, the risk is less. The formula for determining A can be solved for $z_{\alpha/2}$ to give $z_{\alpha/2} = z_{\beta}(M/A - 1)$.

For any achieved precision A, this formula can be used to calculate the effective alpha risk. Whether an auditor selects the positive or negative approach is a matter of personal choice—either approach is valid as long as it is consistently followed. When the desired risks are the same, as in the illustration, the same sample size will result from using either approach. This is verified in the example by observing that the desired standard error of the estimate was the same under either approach. Using the positive approach, the desired precision of $60,500 at an interval reliability of .95 means a desired standard error equal to the precision divided by the corresponding reliability factor (1.96). This gives a desired standard error of $30,867 ($60,500/1.96).

Employing the negative approach, the desired precision of $39,500 at a one-sided reliability of .90 means a desired standard error of $30,859 ($39,500/1.28). The slight difference in the desired standard errors is caused by rounding.

4

Attribute Methods

Attribute methods encompass any sampling situation where a qualitative characteristic is associated with each sampling unit of a population. All attribute methods have one of three basic purposes: estimation, decision, or discovery. All of the methods are directly or indirectly concerned with the percentage of the sampling units that possess the qualitative characteristic known as the *population occurrence rate*. Attribute estimation is appropriate when the objective is to estimate the population occurrence rate within some limits of accuracy as represented by the desired precision and reliability. A decision technique is appropriate when the objective is to decide whether the population occurrence rate exceeds some tolerable limit, and the auditor wants to control the risks of an erroneous decision. Discovery sampling is appropriate when the objective is to observe at least one sample occurrence whenever the population occurrence rate equals or exceeds some critical level.

Various techniques under each of these methods are discussed in this chapter. The discussion of decision techniques covers both samples of predetermined size and sequential samples. Sampling proportional to size (pps sampling) is discussed separately, since it can be used for any of the three basic purposes.

Because attribute methods are usually most appropriate for compliance tests, the illustrations in this chapter are primarily in that area. Attribute methods may also be appropriate for substantive tests, particularly those attributes that have a determinable effect on the monetary amount.

Sampling Unit

When the auditor wants to use any form of attribute sampling, he must carefully define the sampling unit and decide on an appropriate frame. The only requirement in defining the sampling unit is that it must either possess the attribute or not. For example, a sampling unit may be a purchase, a cash disbursement, a sale, an account receivable, an inventory item, or a group of inventory items.

The attribute associated with each sampling unit must be so defined that there is no ambiguity concerning whether or not it occurs. For example, suppose the sampling unit is a disbursement made by a check and the attribute concerns whether a clerk compared the check amount to the amount of the vendor's invoice. The attribute could be defined as "absence of the clerk's initials signifying that the check amount was compared to the vendor's invoice." Using this definition, the presence of the initials would constitute compliance even though the check amount did not agree with the amount on the vendor's invoice. To avoid this situation, the defined attribute could be modified to include the statement "or the check amount does not agree with the amount shown on the vendor's invoice." An occurrence would then correspond to absence of the clerk's initials or presence of the initials but a disagreement between check amount and the vendor's invoice.

Since the required sample size is stated in terms of the number of sampling units that must be examined, it is advantageous to make the sampling unit as small as possible, consistent with the auditor's objectives and the client's accounting system. For example, if the audit objectives can be equally served using a check or an invoice as a sampling unit, and if either is practicable in terms of the client's accounting system, the auditor would be better off choosing the invoice. The auditor should also consider, however, whether a separate selection of checks will be required or whether it will suffice to examine the checks corresponding to the selected invoices. Thus, audit objectives are inseparably intertwined in defining the appropriate sampling unit.

Frame

In many applications of attribute methods, the auditor can select one of several possible frames—each meeting his audit objectives and each being complete or capable of being tested for completeness. In such circumstances, the most convenient frame should be selected. For example, when examining cash disbursements, the cash disbursements book or the check register might be more convenient to use than the voucher register.

An additional consideration in deciding on the frame is the fact that several attributes are usually examined at the same time, which entails selecting a frame that is (1) appropriate for several audit objectives, (2) complete, and (3) convenient to use for each attribute.

Attribute Estimation

When the auditor selects an unrestricted random sample without replacement and determines whether each selected sampling unit possesses the defined attribute, he can use the sample results to estimate the population occurrence rate. In this circumstance the sample occurrence rate is equal to the number of occurrences divided by the sample size. The observed sample occurrence rate is the estimate of the unknown population occurrence rate.

As described in chapter 2, the sampling distribution of the observed number of occurrences in a sample of size n depends upon both the population size (N) and the population occurrence rate. The exact sampling distribution can be closely approximated by the binomial when the sampling fraction (n/N) is less than .10, and by the Poisson when both the sampling fraction and the population occurrence rate are less than .10.[1] When the sampling fraction exceeds .10, a correction factor can be used to improve either approximation.

The sampling distribution has a mean value equal to the population occurrence rate,[2] but the distribution is not symmetric about the mean. The importance of this asymmetry to the auditor is that a precision interval is composed of two parts of unequal length. One part extends from the sample occurrence rate to the upper precision limit, and this part is known as the upper precision. The other part extends from the sample occurrence rate to the lower precision limit and is known as the lower precision. In variables techniques, the upper precision and lower precision have a common value but in attribute estimation, the upper precision exceeds the lower precision whenever the observed sample occurrence rate is less than .50.

Determining Sample Size

Sample size for attribute estimation may be determined either to achieve a desired precision interval at a specified two-sided reliability (R), or to achieve a desired upper precision limit at a specified one-sided reliability (R_1). In either case, a table can be used.[3] Additionally, a computer program such as the one described in chapter 9 (ATS1Z1) may be used.

The procedure for determining the sample size when the desired upper precision limit is specified at a one-sided reliability (R_1) and a table is used as follows:

1. Decide on an anticipated occurrence rate.
2. Find the table corresponding to the specified reliability (R_1).
3. Determine the sample size corresponding to the anticipated occurrence rate and the desired upper precision limit.

The anticipated occurrence rate should represent the auditor's conservative expectation of what he will observe. Since the achieved upper precision limit will exceed the desired upper precision limit whenever the observed occurrence rate exceeds the anticipated rate, it is important that the auditor set the anticipated occurrence rate slightly larger than the rate considered likely based on a preliminary sample, past experience, or other knowledge.

For example, assume the auditor desires an upper precision limit of .05 at .95 one-sided reliability and sets the anticipated occurrence rate at .02 even though a rate of .01 seems likely. Referring to an appropriate table the auditor finds that the required sample size in these circumstances is about 180 sampling units. For a sample of this size, the achieved upper precision will not exceed .05 as long as the sample occurrence rate is less than or equal to .02. Consequently, the desired upper precision limit will be achieved whenever the observed number of occurrences does not exceed four. Had the auditor set the anticipated occurrence rate at .01 instead, the sample size would have been only about 100 units, but the desired upper precision limit would have been achieved only for an observed occurrence rate of up to

1. Because of typically small population occurrence rates in auditing applications, the normal approximation is not considered appropriate for attribute methods.
2. This is another way of stating that the sample occurrence rate is an *unbiased* estimate of the population occurrence rate.
3. Tables for computing sample sizes are presented in AICPA [2], vol. 6.

.01 (not more than one occurrence). The additional 80 units provide a safety factor for mistakes in what the auditor thinks is the likely occurrence rate.

A similar procedure can be used to determine the sample size required when the auditor specifies a desired precision at a two-sided reliability R. Since the upper precision is larger than the lower precision, determining a sample size that yields a desired upper precision will automatically yield a precision interval that meets the desired specification.[4] As an illustration, if the auditor desires a precision of .03, then using .03 as the desired upper precision means that the desired lower precision is smaller than .03.

When the auditor specifies a desired upper precision at a two-sided reliability (R), but has only the tables designed for a desired upper precision limit at a one-sided reliability (R_1), it is necessary to (1) convert the desired upper precision to a desired upper precision limit, and (2) convert the two-sided reliability (R) to the corresponding one-sided reliability (R_1).

The first conversion is done by adding the desired upper precision to the anticipated occurrence rate. For example, if the anticipated occurrence rate is .02, and the desired upper precision is .03, then the desired upper precision limit is .05 (.02 + .03). Deciding the appropriate value of the anticipated occurrence rate involves exercising the same degree of conservatism suggested previously.

When the desired reliability is specified as R for a precision interval, the following relationship converts this to the equivalent of one-sided reliability (R_1):

$$R_1 = \frac{R + 1.0}{2}.$$

Thus, when R is designated as .90, for example, the corresponding one-sided reliability (R_1) equals .95.

After making these conversions, the appropriate sample size can be determined by following the same procedures previously set forth for determining the sample size on the basis of a desired upper precision limit at a specified one-sided reliability (R_1). For example, when the auditor sets the anticipated occurrence rate at .02 and desires an upper precision of .03 at .90 two-sided reliability, the conversions illustrated above allow use of tables for an anticipated occurrence rate of .02 and a desired upper precision limit of .05 (.02 + .03) at a .95 one-sided reliability (R_1). As previously illustrated, those tables indicate a required sample size of about 180 units.

Evaluating Results

After selecting the sample and completing the audit procedures on each selected sampling unit, the auditor can evaluate the results statistically. As previously indicated, the estimated population occurrence rate equals the observed number of occurrences divided by the sample size.

Tables can be used to determine the achieved upper precision limit at a one-sided reliability of R_1.[5] In addition, a time sharing computer program such as the one described in chapter 9 will compute this for any sample size. The results from either source is the estimate together with the achieved upper precision limit at a one-sided reliability of R_1. The statistical interpretation of this is that, repeatedly following the

4. This applied when the observed occurrence rates are less than .50. For situations where the rates exceed .50, the lower precision is larger than the upper precision.
5. See the tables in AICPA [2], vols. 2 and 6.

same procedures, the auditor would expect the resulting upper precision limit to equal or exceed the population occurrence rate $100R_1$ percent of the time.

For example, one observed occurrence in an unrestricted random sample of 120 would indicate an achieved upper precision limit of .04 at a .95 one-sided reliability level. The auditor may be tempted to state that the probability is .95 that the population occurrence rate does not exceed .04—but this would be an unwarranted assertion since the probability concept used measures only frequency. For a given population, the true but unknown occurrence rate must be less than, equal to, or greater than .04. From the frequency viewpoint, the auditor can assert that were he to choose unrestricted random samples of 120 repeatedly and calculate the achieved upper precision limit at the .95 one-sided reliability, about 95 percent of such achieved upper precision limits would equal or exceed the population occurrence rate. This places the emphasis on the procedure and the reliability refers to the success ratio of applying that procedure a large number of times.

When the auditor desires a precision interval, he must determine a lower precision limit in addition to the upper precision limit. This can be done using either tables or the same computer program mentioned previously.[6] When the tables employ one-sided reliability (R_1) the conversion to interval reliability must be done. This is accomplished using the following equation:

$$R = 2R_1 - 1.0.$$

Thus, when R_1 is specified as .95, for example, the corresponding interval reliability equals .90.

For example, four observed occurrences in an unrestricted random sample of 120 would indicate an achieved upper precision limit of .08 at a one-sided reliability of .95 and a lower precision limit of .01 at a one-sided reliability of .95. The precision interval ranges from .01 to .08 at a reliability of .90. Notice that the observed occurrence rate is $4/120 = .033$. Therefore, the achieved upper precision is .047 ($.08 - .033$) and the achieved lower precision is smaller at .023 ($.033 - .01$).

The auditor can report that the achieved precision interval resulted from following a procedure that produces intervals containing the population occurrence rate with a frequency of $100R$ percent.

Attribute Decision

The results from an unrestricted random sample selected without replacement can be used in deciding whether a population occurrence rate is as high or higher than some specified rate.[7] This attribute method can be used in deciding whether compliance with pertinent internal accounting control procedures justifies the auditor's planned reliance. Deciding whether compliance is satisfactory requires a standard and was discussed in chapter 1. The standard is expressed as a particular threshold occurrence rate, labeled P_0, so that were compliance deviations as large as

6. Tables for this purpose are included in AICPA [2], vol. 2.
7. This topic is often discussed under *acceptance sampling,* which was developed to help decide when a manufactured lot of items is satisfactory. See Arkin [4], Cyert and Davidson [6], and Vance and Neter [21]. Since the jargon of acceptance sampling seems ill-suited to auditing, the topic is developed here without the usual references to that material.

this rate, the auditor would adjust his substantive tests to reflect less reliance than was planned on that accounting control or set of controls. On the other hand, compliance deviations at any rate smaller than P_0 would be consistent with the auditor's planned reliance.

Fixed Sample-Size Plans

To decide on the basis of a sample of size *n* whether the occurrence rate may be as large as P_0, the auditor needs to specify a decision rule. The rule stipulates the action to be taken for each possible observed number of occurrences. In particular, the auditor specifies a critical number of occurrences (or equivalently a critical sample occurrence rate), so that when the observed number of occurrences exceeds the critical number of occurrences (or the sample occurrence rate exceeds the critical sample occurrence rate), he decides that the population occurrence rate may be as large as P_0. When the observed number of occurrences is less than or equal to the critical number of occurrences, he decides that the occurrence rate is smaller than P_0.

For example, suppose the auditor judges that compliance would not be satisfactory from the standpoint of planned reliance if the occurrence rate is as large as 5 percent.[8] It must then be determined what sample size and what number or rate of observed occurrences would lead to deciding that the occurrence rate is as large as 5 percent.

Instead of specifying the critical number of occurrences, the auditor may specify a desired upper precision limit equal to P_0. When the achieved upper precision limit exceeds P_0, the auditor decides that the occurrence rate may be as large as P_0 and when the achieved upper precision limit is not larger than P_0, he decides that the occurrence rate is smaller than P_0. That these decision rules are equivalent can be easily seen by examining a table for evaluating attribute estimation samples employing one-sided reliability (R_1). For example, at .95 one-sided reliability, a sample of 120 units gives an achieved upper precision limit of .06 when two occurrences are observed, and an achieved upper precision limit of .07 when three occurrences are observed. Consequently, it makes no difference whether the decision rule is stated in terms of the observed number of occurrences—decide that the occurrence rate is as large as 6 percent when more than 2 occurrences are observed; or in terms of the achieved upper precision limit—decide the occurrence rate is as large as 6 percent when the achieved upper precision limit exceeds 6 percent.[9]

With any decision rule, two types of errors are possible: (1) incorrectly deciding the population occurrence rate is smaller than, say, 6 percent when it is as large or larger, and (2) incorrectly deciding the occurrence rate is as large as 6 percent when it is smaller. Both these errors may be controlled by using an appropriate sample size and designating an appropriate critical number or rate of occurrences.

The more serious error is relying on an accounting control when the reliance is not justified.[10] This type of error, denoted by delta (δ), is measured by the complement of the one-sided reliability (R_1) when the population occurrence rate is P_0 ($\delta = 1 - R_1$). In the previous illustration, the one-sided reliability of .95 means that there is a 5 percent risk of incorrectly deciding that the population occurrence rate is less than .06

8. See the discussion in chapter 1 for a suggested way of arriving at this judgment.
9. Table used is from AICPA [2], vol. 2.
10. This risk is called the *risk of unwarranted reliance*. It is called the beta risk in acceptance sampling (see [21]), but in this book the symbol "δ" will be used instead.

when, in fact, it equals .06. This is the maximum risk in the sense that for any population occurrence rate larger than .06, the risk is smaller than 5 percent.

The fact that the risk grows smaller as the population occurrence rate increases beyond P_0 is very important in practice. For example, in a sample of 120, the auditor would have a 5 percent risk if adopting the rule of deciding the population occurrence rate is as large as .06 whenever the achieved upper precision limit exceeds .06 (or the observed number of occurrences exceeds 2). This risk decreases to less then 1 percent when the population occurrence rate is .07 and is nearly zero (.001) when the population occurrence rate is .09.

Selecting an appropriate value for the threshold rate, P_0, is not an easy task. Following the suggestion made in chapter 1, the auditor could designate P_0 as the rate corresponding to an expected potential material amount of monetary error. The expected potential monetary error associated with a particular rate of compliance deviation equals the rate times the number of transactions times the average dollar amount of each transaction or, equivalently, equals the rate times the total dollar amount of the transactions.[11]

It is not necessary to determine the threshold rate (P_0) exactly, because the probability of deciding that compliance is satisfactory changes gradually as the population occurrence rate moves away from P_0. For example, in the illustration a sample of 120 afforded a risk of unwarranted reliance (δ) equal to .05 when $P_0 = .06$. The same sample size affords a probability of .06 of reaching the conclusion that compliance is satisfactory when the population occurrence rate is .05. Consequently, the auditor would have nearly the same risk of unwarranted reliance as long as the threshold rate was set somewhere between .05 and .06. The exact point selected makes little difference.

The other type of error—deciding that the occurrence rate is as large as P_0 when in fact it is smaller—is not as serious as the first type discussed, but it is important.[12] Unnecessarily increasing the audit work beyond that consistent with the planned reliance increases the cost of auditing, and hence this type of error also needs to be controlled. To exercise control, the auditor may designate an occurrence rate P_1 so that whenever the occurrence rate is as small as P_1, the risk of incorrectly deciding that the occurrence rate is as large as P_0 is equal to some specified value. For example, if the auditor designates P_1 as .01 in the sample of 120, the risk of deciding that the occurrence rate is as large as .06 when in fact the population occurrence rate is as small as .01, is equal to about .12.

Again, in deciding on P_1, the auditor is attempting to determine a population occurrence rate that is consistent with controlling the risk of overauditing when there is excellent compliance. The risk of overauditing diminishes for population occurrence rates smaller than P_1, and gradually increases as the occurrence rate increases beyond P_1.[13] Operationally, the auditor might ordinarily select P_1 in the range of .005 to .01.

---

11. This expected potential monetary error may be useful whenever each transaction has the same chance of containing a compliance deviation. If the auditor knows that some transactions are more likely to contain evidence of compliance deviations, he should incorporate this information into his planning.
12. This risk is called the *risk of overauditing*. In acceptance sampling terminology, this is known as the alpha risk (see Vance and Neter [21]). No special symbol is used in this book.
13. In practice, specifying P_0 and P_1 as far apart as is reasonable results in the smallest sample size. When the population occurrence rate is zero, of course, there is no risk of overauditing resulting from the sampling process.

To summarize, the auditor desiring to decide whether an occurrence rate is as large as some specified value P_0, determines a sample size and a decision rule based on whether the achieved upper precision limit exceeds P_0 or not. The risk of unwarranted reliance (δ) is equal to the complement of the one-sided reliability, R_1. The risk of overauditing may be controlled at a designated occurrence rate of P_1 by determining an appropriate sample size.

Determining Sample Size. The sample size is determined by the four quantities P_0, P_1, the risk of unwarranted reliance (δ), and the risk of overauditing. In some situations the auditor may omit from consideration P_1 and the associated risk of overauditing. When only P_0 and the risk of unwarranted reliance are considered, the auditor can determine the sample size using exactly the same procedure as in attribute estimation but with P_0 corresponding to the desired upper precision limit and the complement of the desired risk of unwarranted reliance corresponding to the one-sided reliability (R_1). To repeat, these steps are as follows:

1. Decide on an anticipated occurrence rate.
2. Find the table corresponding to the one-sided reliability (R_1).
3. Look up the sample size corresponding to the anticipated occurrence rate and the desired upper precision limit (P_0).

Following this procedure, the auditor's anticipated occurrence rate equals the critical sample occurrence rate, so that whenever he observes more than the critical number of occurrences, the sample evaluation will indicate that the population occurrence rate may exceed P_0. For example, using an anticipated occurrence rate of .01, and a desired upper precision limit of .05, with a one-sided reliability of .95, the required sample size is about 100.[14] After auditing the 100 selected sampling units, the sample evaluation will indicate that the occurrence rate may be as large as .05 whenever more than one ($100 \times .01$) occurrence is observed and that the population occurrence rate is less than .05 whenever either zero or one occurrences are observed. The risk of unwarranted reliance is .05, or the complement of the one-sided reliability.

The deficiency in using this procedure is that the risk of overauditing is unknown and uncontrolled whenever the population occurrence rate is not zero. For example, the risk of overauditing is .26 when the population occurrence rate is .01, and this risk increases to nearly .60 when the occurrence rate is .02. This means that the auditor will needlessly place less reliance on the control about one time in four whenever the population occurrence rate merely equals the anticipated occurrence rate.

To prevent this, the sample size and the critical number of occurrences can be determined to control the risk of overauditing when the population occurrence rate equals a specified P_1. For example, if the auditor specifies a .05 risk of overauditing at $P_1 = .01$ and a .05 risk of unwarranted reliance at $P_0 = .05$, the required sample size is about 180 and the critical number of occurrences is four.

The required sample size may be determined by using tables of the cumulative binomial distribution (see Aiken [1]) or a computer program such as one presented in chapter 9. Using tables, the procedure is to locate a sample size and critical number so that the probabilities of exceeding the critical number are the specified risk of

14. See the tables in AICPA [2], vol. 6.

overauditing at P_1 and the complement of the specified risk of unwarranted reliance at P_0. This can be done in the following way:

1. For a trial sample size, determine the critical number corresponding to the specified risk of unwarranted reliance at P_0.
2. Determine the probability of exceeding that critical number at P_1. If this exceeds the desired risk of overauditing, increase the sample size sufficiently to increase the critical number by one and repeat.

To illustrate, again consider the example of a .05 risk of overauditing at $P_1 = .01$ and a .05 risk of unwarranted reliance at $P_0 = .05$. The following table shows the results of the described procedures.

Trial Sample Size	Critical Number	Risk of Unwarranted Reliance at $P_1 = .05$	Risk of Overauditing at $P_0 = .01$
100	1	.037	.264
130	2	.039	.142
160	3	.039	.078
180	4	.051	.036

As the example shows, the cost of controlling the risk of overauditing at $P_1 = .01$ is increasing the sample size from 100 to 180. The benefit is to decrease the risk from about .26 to .04 that the auditor would needlessly extend his planned substantive procedures. Unless the benefit exceeds the additional cost, the sample size for the compliance test should not be increased. This aspect of planning will be discussed more thoroughly in chapter 7.

Evaluating Results. The evaluation of unrestricted random samples selected without replacement to decide whether the occurrence rate is as high as P_0 is straightforward. After performing the audit procedures on the selected sampling units, the auditor determines the number of observed occurrences. If the number of observed occurrences exceeds the critical number of occurrences or if the achieved upper precision limit exceeds P_0, the auditor decides the population occurrence rate may be as large as P_0. Otherwise, the decision is that the population occurrence rate is smaller than P_0 and is therefore acceptable for the purpose of placing the planned reliance on the pertinent accounting control.

Following this procedure the auditor has a determined risk of incorrectly deciding the occurrence rate is smaller than P_0 for any population occurrence rate. In particular, the risk of unwarranted reliance is equal to the complement of the one-sided reliability (R_1) when the occurrence rate equals P_0.

The risk of overauditing may likewise be determined for any specified population occurrence rate (P_1).

Sequential Plans

In this section, sequentially determined sample sizes for making a decision concerning a population occurrence rate are discussed. The reason for using a sequential sample rather than determining a fixed sample size in advance is that many times fewer sampling units will be required before a decision is reached. More specifically, a sequential sample will require fewer observations whenever the population occurrence rate is either very low or much higher than anticipated.

At each stage of the sampling process, the sample evidence is evaluated to determine whether a decision concerning the population occurrence rate can be reached or whether additional sample information is required. Sequential decision plans for attributes involve specifying an initial sample size, an incremental sample size to be added at each stage of the sampling process, and the total number of stages possible.

A class of sequential plans with four stages has been developed and tested in actual situations. These plans have proved useful in reducing the number of required observations compared to using a predetermined sample size with the same risks of error. In addition, the plans are quite simple. In the first stage an initial sample of size n_I is audited and if additional observations are required, the same number (Δn) of additional observations is added at each of three possible subsequent stages. Both the initial sample size (n_I) and the incremental sample size (Δn) are chosen to limit the risk of unwarranted reliance and the risk of overauditing as specified by the auditor.

Sampling Distribution. The auditor may reasonably ask why it is necessary to design specific sequential plans rather than construct them based on attribute tables designed for predetermined sample sizes. The answer is that sequential plans constructed from standard attribute tables have a larger risk of unwarranted reliance than the nominal risk based on the tables. To illustrate this, consider the following sequential plan (sometimes called *stop-or-go sampling*) designed to decide whether the population occurrence rate is as great as .08, with a desired risk of unwarranted reliance equal to .10. In constructing the sampling plan, the desired upper precision limit is .08 and the one-sided reliability (R_1) is .90.

	Sample Size		Decide population occurrence rate is less than .08 when number of observed occurrences is less than or equal to	Decide population occurrence rate may be as great as .08 when number of observed occurrences is at least
Stage	Incremental	Cumulative		
1	50	50	1	8
2	20	70	2	8
3	30	100	4	8
4	20	120	5	8
5	30	150	7	8

At each cumulative sample size, the number of observed occurrences yielding an upper precision limit of .08 at a one-sided reliability of .90 was determined. In a sample of 100, for example, four observed occurrences yield an upper precision limit of .08 at .90 one-sided reliability. For the above stop-or-go plan the largest sample size of 150 was designed to be three times the initial sample size.[15] Since observing eight occurrences in a sample of 150 yields an upper precision limit exceeding .08 at .90 one-sided reliability, observing at least that number of occurrences at any stage would lead to the decision that the population occurrence rate could exceed .08.

Unfortunately, the foregoing plan does not have a risk of unwarranted reliance equal to .10 (the complement of the nominal one-sided reliability) but one of nearly .16. The increase over the nominal risk occurs because the sample evidence is used at each stage to decide whether to stop or continue to the next stage. To develop a sequential plan that maintains this risk at the desired level, consideration must be given to the effect that the intermediate sample results have on determining the final total sample size. This has been done for the plan described below.

Determining Sample Size. To determine the initial sample size (n_I) and the incremental sample size (Δn) the auditor specifies the same four quantities needed when the total sample size is predetermined—P_0, P_1, the risk of unwarranted reliance at P_0, and the risk of overauditing at P_1. To minimize the required number of tables under this sequential plan, the tables in Appendix 3C allow the auditor to specify only P_1 as either .005 or .01, for which the risk of overauditing does not exceed .05 or .10, respectively. The table then shows the initial sample size and incremental sample size corresponding to a specified P_0 and a specified risk of unwarranted reliance. The numbers in the table corresponding to a specified risk of unwarranted reliance (δ) and a specified threshold rate (P_0) represent the initial sample size (n_I). The incremental sample size (Δn) corresponding to a threshold rate (P_0) is the same for each risk level and is shown on the last line of the table.

Evaluating Results. The following indicates the sequential procedure to be followed. The first stage calls for examining n_I randomly selected sampling units. When the observed number of occurrences in the first stage is zero, sampling stops and the auditor decides that the population occurrence rate is smaller than P_0. On the other hand, if four or more occurrences are observed, sampling stops and the appropriate decision is that the population occurrence rate may be as large as P_0. When the observed number of occurrences is one, two, or three—that is, more than zero but less than four—an additional sample of Δn sampling units is selected.

When the number of occurrences in the cumulative sample ($n_I + \Delta n$) of the second stage is one, sampling stops and the auditor decides that the population occurrence rate is less than P_0. If the number of observed occurrences is four or more, sampling stops and the auditor decides that the population occurrence rate may be as large as P_0. When the number of observed occurrences at the second stage exceeds one but is smaller than four, the auditor goes to the third stage.

At the third stage sampling stops when the number of occurrences equals two, in which case the auditor decides that the population occurrence rate is smaller than P_0; or, sampling stops when the number equals or exceeds four, in which case the auditor decides that the population occurrence rate may be as large as P_0. When the number of observed occurrences is three, the fourth and last stage sample is selected and evaluated.

When the final sample has three occurrences, the auditor decides that the population occurrence rate is smaller than P_0, and when it has four or more, the auditor

15. This conforms to the suggested design of stop-or-go sampling plans. See Wilburn [23].

decides that the population occurrence rate could equal or exceed P_0. This procedure is summarized in the following table.

Stage	Sample Size: Incremental	Sample Size: Cumulative	Decide population occurrence rate is less than P_0 when number of observed occurrences equals	Decide population occurrence rate may be as great as P_0 when number of observed occurrences is at least
1	n_I	n_I	0	4
2	Δn	$n_I + \Delta n$	1	4
3	Δn	$n_I + 2\Delta n$	2	4
4	Δn	$n_I + 3\Delta n$	3	4

The procedure is designed to achieve the auditor's specified risk of unwarranted reliance, which is equal to the probability that the auditor incorrectly decides that the population occurrence rate is smaller than P_0 when it actually equals P_0, as specified by the auditor.

The risk of overauditing can also be controlled by using the tables. This risk is no larger than .05 when the table corresponding to $P_1 = .005$ is used and no larger than .10 when the table corresponding to $P_1 = .01$ is used. The actual value of the risk is generally smaller than these bounds.

Curtailed Sampling

When the auditor wants to decide whether the population occurrence rate is as large as P_0 and wants to limit the number of observations to as few as possible to achieve a specified risk of unwarranted reliance at P_0 and a specified risk of over-auditing of P_1, a sequential procedure should be used, such as one suggested in the previous section.[16] However, when this is not practicable, the auditor may gain some of the benefits of the sequential approach by stopping sampling whenever the number of occurrences exceeds the number corresponding to an achieved upper precision limit of P_0.

This technique, called *curtailed sampling,* reduces the predetermined sample size only when there is an early indication that the sample is going to result in the auditor's deciding that the occurrence rate is P_0 or larger. In order to decide that the occurrence rate is less than P_0, all the selected sampling units must be examined. For example, suppose that the auditor determines that a sample of 150 is needed to decide whether the population occurrence rate is equal to or larger than .06, with a specified risk of unwarranted reliance equal to .05. Examining the attribute tables[17] shows that the sample will lead to deciding that the population occurrence rate is smaller than .06 only when the number of observed occurrences is four or fewer. Consequently, the auditor can adopt the rule of stopping the observations as soon as

16. There are many such sequential plans that could be adopted, ranging from the ordinary sequential probability ratio test to plans with a fixed number of stages.
17. See Table 3 in AICPA [2], vol. 6.

five occurrences are observed. Using this curtailed sample, he examines fewer than the entire 150 sampled items whenever the observed results indicate the population occurrence rate could be as large as .06. On the other hand, the only way the auditor can decide that the population occurrence rate is less than .06 is to examine all 150 sample items.

The only disadvantage of using curtailed sampling is that when the sampling is stopped before the entire sample has been examined, the fraction of observed occurrences does not provide an estimate of the population occurrence rate. This, however, seems a minor disadvantage when the purpose of sampling is to decide whether the occurrence rate is below P_0.

Discovery Sampling

An unrestricted random sample may also be used for the purpose of discovering examples of certain attributes. The chance of discovery, of course, is affected by the number of occurrences existing in the population. As might be expected, achieving a discovery objective is easy when there are numerous occurrences in the population but is nearly impossible if the occurrences are very rare. In auditing, discovery sampling (sometimes called *exploratory sampling*) is often used when the auditor believes the population occurrence rate is near zero but wants to examine enough sampling units so that in case he is incorrect, he will observe at least one occurrence.[18] The observation of one occurrence allows him to decide that the population occurrence rate is not zero—and this is often sufficient to indicate what other course of action he should take.

Sampling Distribution

For discovery samples using unrestricted random sampling without replacement, the correct sampling distribution is the hypergeometric distribution. Moreover, since the only event of concern is the probability of at least one occurrence in a sample of a given size, the exact probability can be easily determined using a very simple computer program.[19] The probability depends upon the population size N and the number or rate of occurrences in the population.

When the sampling fraction (n/N) is small, the probability of observing at least one occurrence in a sample of n units can be easily approximated. The desired probability is the complement of the probability that no occurrences are observed. The probability that no occurrences are observed in a sample of size n is approximately equal to the complement of the population occurrence rate raised to the nth power.

Determining Sample Size

The sample size required to observe at least one occurrence can be determined for any population size by specifying (1) the population occurrence rate (P_0) for which it is desired that at least one occurrence be observed and (2) the desired probability of observing at least one occurrence (the probability of discovery) at that rate (P_0).

---

18. This is closely related to the fact that when the auditor has no information concerning which units are likely to represent an occurrence, the strategy which maximizes the probability of observing an occurrence is to take an unrestricted random sample.
19. See chapter 9 for a description of such a program.

In specifying P_0, the auditor should ask how large the population occurrence rate would have to be before needing a specified probability of observing at least one occurrence. Since, for most applications, the auditor anticipates that the occurrence rate is close to zero, P_0 should be selected to represent the smallest population occurrence rate that would cause him to take an action different from what he had planned. In this case, P_0 represents a *threshold occurrence rate.*

The probability of discovery at the threshold occurrence rate P_0 is the second quantity that must be specified. The probability of observing at least one occurrence will naturally exceed this specified probability of discovery whenever the population occurrence rate exceeds P_0. Likewise, the actual probability will be smaller than the specified probability whenever the population occurrence rate is smaller than P_0.

Using the values P_0 and the desired probability of discovery together with the population size (N), the auditor can determine the appropriate sample size by (1) using a time sharing computer program, (2) using the approximate formula with a calculator, or (3) looking up the sample size in a table.[20] In case the auditor uses a calculator, the effect, if any, of the population size is often neglected for computational convenience. Some tables reflect the population size (those based on the hypergeometric) while others do not. Among the latter, a very convenient table is based on the Poisson approximation—this can be found in volume 6 of [2].

To illustrate, suppose the auditor has a population of 10,000 for which he wants the probability to be .95 of observing at least one occurrence when the population occurrence rate is .01. In other words, if there are 100 ($.01 \times 10{,}000$) occurrences in the population, he wants the probability of observing one or more to be .95. Examining a table based on the hypergeometric distribution the auditor finds the required sample size is 300. Since this is only 3 percent of the population, the binomial or Poisson approximations will yield the same answer.

Evaluating Results

When an occurrence is observed, the objective of the discovery sample has been satisfied and no further statistical evaluation is necessary. When no occurrence is observed, the sample may be evaluated as any unrestricted random sample with zero occurrences. This means that even though the sample was not selected for the purpose of estimating a population occurrence rate, or for deciding whether the population occurrence rate is as great as P_0, evaluation for either of these purposes is valid. In making such an evaluation, the threshold rate P_0 becomes the achieved upper precision limit and the probability of discovery becomes the one-sided reliability (R_1). This, of course, means that with zero occurrences, the appropriate decision is that the population occurrence rate is less than P_0, and the risk of unwarranted reliance is the complement of the one-sided reliability ($1 - R_1$).

In addition to using discovery sampling in connection with compliance tests, it can be used in connection with substantive tests as will be discussed more fully in chapters 5 and 6. When used in the substantive area, the population occurrence rate refers to the fraction of the sampling units that have a non-zero difference between the audited amount and the recorded amount. The sample size would be chosen to afford a specified probability of discovery in terms of the probability of observing at least one difference when the fraction of difference in the population is at the threshold level (P_0).

---

20. Tables for this purpose are in AICPA [2], vol. 2.

When no difference is observed in the sample, the auditor will conclude that the fraction of population items with any monetary difference is less than P_0. If, in addition, the auditor can establish the largest amount of difference that is possible, he can compute a conservative but possibly useful bound on the total monetary difference by extending that largest amount by the achieved upper precision limit, which equals P_0 when no differences are observed. For example, suppose that 300 requests for confirmation selected at random from 5000 accounts receivable exhibit no differences between the audited amount and the recorded amount. Then at a one-sided reliability of .95 the auditor concludes that no more than 1 percent or 50 (5000 × .01) of the accounts could be misstated by any amount. If the largest recorded balance is $2000, then that amount constitutes the largest amount of overstatement possible for any one account. The statistical evidence, therefore, indicates that accounts receivable could not be overstated by more than $100,000 (50 × $2000) with a risk of being incorrect equal to .05 or the complement of the .95 one-sided reliability.

The limit to monetary error obtained in this manner is, of course, quite conservative. Other, less conservative methods may also prove useful to the auditor. For example, in the above example, the auditor could determine the recorded amount of the 50 largest accounts not included in the sample, and use the resulting total as an upper limit of overstatement at a .95 one-sided reliability.

Sampling Proportional to Size

In testing compliance with pertinent accounting controls, the auditor may be interested in establishing some relationship between the incidence of compliance deviations and the monetary amount affected by a particular type of deviation. One method of doing this is to select transactions with probability proportional to size—using the recorded dollar magnitude of the sampling unit as the measure of size and one of the pps methods described in chapter 2. For each selected transaction, the auditor establishes whether the defined attribute is present just as in an ordinary attributes sample. The total number of sample occurrences divided by the sample size provides an estimate of the fraction of the total recorded dollar amount that is associated with compliance deviations.[21] Multiplying that estimate by the total recorded dollar amount yields an estimate of the total dollars associated with the defined attribute.

The sampling distribution of the sample occurrence rate (number of sample occurrences divided by the sample size) is approximated by the binomial distribution and consequently, tables based on the binomial distribution or the Poisson distribution may be used to determine sample sizes and evaluate the sample results for either estimating or decision-making purposes.

To illustrate, suppose that from a population of payments made by check a pps sample is to be selected and examined for the presence of several attributes. One of the attributes is defined as "payment was not properly authorized." The population of payments covers a period of ten months and has a recorded amount equal to $2 million. The auditor decides that his planned reliance on this control will be justified if, at a reliability of .95, no more than .05 or $100,000 of the recorded amount has been paid without proper authorization. Past experience leads to anticipating few if any nonauthorized payments, so the auditor designates the anticipated occur-

21. See Appendix 6B for a technical discussion of this procedure.

rence rate as .01 or $20,000. Examining an attribute table, the required sample size is determined to be 100.[22]

To evaluate the sample results, the auditor will determine the number of the selected checks that were issued without proper authorization. From the attribute table, the auditor notes that if, at most, one of the 100 checks examined has not been properly authorized, it can be concluded that the total dollars paid without proper authorization does not exceed $100,000 at a .95 one-sided reliability. Consequently, if the evidence is used in decision-making, the auditor will decide that the planned reliance is justified when, at most, one of the sampled checks has not been properly authorized. The risk of unwarranted reliance is .05 or the complement of the one-sided reliability.[23]

22. See tables in AICPA [2], vol. 6.

23. In the illustration, the risk of overauditing has not been explicitly considered. Using this sample size, there is a risk of .26 of overauditing (observing more than one unauthorized payment) when .01 of the check amounts ($20,000) have been unauthorized.

5

Unstratified Variable Methods

When a variable, usually a dollar amount, is associated with each sampling unit, the auditor may use one of several possible variable sampling methods. These include mean, difference, ratio, and regression estimation. In this chapter, each of these methods is discussed when unstratified sampling is used. The discussion includes (1) conditions for the method to be valid and efficient, (2) ways of determining the required sample size, and (3) evaluation of results.

Chapter 6 discusses the variable sampling methods when stratified sampling is used. Stratified sampling is used in most auditing applications of variable methods. There are, of course, some situations where unstratified sampling may be appropriate, either because there is no practical way to stratify the population or because the population dollar amounts are within reasonably narrow limits. But such situations are rare. Whether the sample is stratified or not, the auditor selects any items that are individually significant on a 100-percent basis.

Why then discuss the unstratified versions of the variable sampling methods? Simply because it will be easier for readers to understand the stratified variable methods if they first understand the methods in the unstratified situation. Variable methods are more complex than attributes. This chapter and the following one have many formulas. These formulas are necessary to express the complex relationships, but fortunately the set of computer programs described in chapter 9 make it unnecessary for the auditor to use these formulas in practical applications.

Mean Estimation

Unstratified mean estimation can be used whenever

1. There is no recorded amount for each sampling unit.
2. The distribution of audited amounts is not highly skewed.[1]

---

1. See Appendix 1 for a definition of *skewness*.

When no recorded amounts exist for the sampling units, mean estimation based on audited amounts is the only applicable method.

When recorded amounts do exist, two circumstances may also lead the auditor to choose mean estimation instead of one of the other variable methods:

1. The general ledger control amount does not agree with the total of the individual recorded amounts.
2. The agreement between the individual recorded amounts and audited amounts is so poor that the correlation between them is less than one-half.[2]

If the auditor cannot establish agreement between the general ledger control amount and the total individual recorded amounts mean estimation should be used. Low correlation makes mean estimation more efficient than any other method except regression estimation.[3]

Efficiency. When recorded amounts exist and are reasonably well correlated (correlation greater than one-half) with the audited amounts, mean estimation of the total audited amount is less efficient than the difference or ratio methods that use the recorded amounts. Similarly, mean estimation is less efficient than regression estimation whenever the correlation is not zero. By ignoring the recorded amounts in such situations, mean estimation can produce results that have little information value to the auditor. For example, suppose an unrestricted random sample of accounts receivable with a total recorded amount of $1 million yielded a $900,000 mean estimate of the total audited amount. If no differences between recorded and audited amounts were observed in the sample (that is, the sample exhibited perfect correlation), this $100,000 discrepancy would be meaningless to the auditor. The lack of any observed differences means that the entire $100,000 discrepancy is caused by sampling variation and by itself does not provide any information about the reasonableness of the recorded amount.

A lack of skewness in the audited amounts is another desirable condition for using unstratified mean estimation. When skewness is present, as is often the case with accounting populations, the sample size required to obtain a good estimate of the total audited amount and to insure that the sampling distribution is approximately a normal distribution may be too large to be practicable. In such a situation, the auditor would ordinarily try to stratify the sampling units.

Sampling Distribution. An unrestricted random sample is selected and an audited amount established for each sample unit. The mean estimate of the total audited amount is formed by multiplying the sample average audited amount by the number of units in the population. In symbols, this may be expressed as

$$\hat{X}_M = N\bar{x}, \quad \text{where } \bar{x} = \frac{\sum x_j}{n}$$

is the sample average audited amount. The symbol $\hat{X}_M$ represents the mean estimate of the total audited amount and N the population size.[4]

2. See Appendix 1 for a definition of *correlation.*
3. See Appendix 1 for a definition of *statistical efficiency.*
4. The subscript M of $\hat{X}_M$ denotes the estimated audited amount is based on the *mean* estimator.

The sampling distribution describes how the values of the mean estimator vary over all the different possible samples. The mean of the sampling distribution is the total population audited amount. Another way of expressing this is to say that the mean estimator is an *unbiased* estimator of the total audited amount.[5]

The standard error of the mean estimator equals the standard deviation of the sampling distribution. This standard deviation equals the standard deviation of audited amounts divided by the square root of the sample size and multiplied by the product of the population size and the finite population correction factor. Unfortunately, the standard deviation of the population of audited amounts is generally not known, and consequently must be estimated from the sample.

The estimated standard deviation of the population of audited amounts is denoted by the symbol S_X. It can be calculated using the following formula:

$$S_X = \sqrt{\frac{\sum x_j^2 - n\bar{x}^2}{n - 1}}.$$

Using this formula to calculate the estimated standard deviation entails performing the following operations.

1. Compute the sum of squares of the sample audited amounts.
2. From that sum, subtract the product of the sample size times the square of the sample mean of the audited amounts.
3. Divide the remainder by one less than the sample size.
4. Take the square root.

The estimated standard error of the mean estimator is then equal to the estimated standard deviation of audited amounts (S_X), divided by the square root of the sample size ($\sqrt{n}$), and multiplied by the product of the population size and the finite population correction factor ($N\sqrt{1 - n/N}$). This estimate is denoted by $\hat{\sigma}(\hat{X}_M)$ (read the estimated standard error of the estimated total audited amount) and is computed by the following formula:

$$\hat{\sigma}(\hat{X}_M) = \frac{NS_X\sqrt{1 - n/N}}{\sqrt{n}}.$$

Statistical theory says that the mean estimator will be approximately normally distributed for sufficiently large sample sizes. The sample size required to achieve a satisfactory approximation to the normal distribution depends upon the shape of the population distribution of audited amounts. Relatively small sample sizes will suffice when the population distribution is nearly symmetric about its mean. Much larger sample sizes are required when the population of audited amounts is highly skewed. Empirical studies of a few accounting populations with only moderate skewness suggest that sample sizes of at least 200 produce satisfactory results.[6]

Determining Sample Size. Sample size should be determined to assure the validity of the normal approximation and to provide the desired precision at the speci-

---

5. See chapter 11.
6. Neter and Loebbecke [17].

fied reliability. In most cases the sample size required to achieve the desired precision would be more than adequate for the normal approximation to be appropriate. The following equation for n expresses the sample size required to give a specified absolute precision of A at an interval reliability of R:

$$n = \frac{N^2 U_R{}^2 S_x{}^2}{A^2 + N U_R{}^2 S_x{}^2}.$$

Basically, this formula for the sample size is derived from the previously cited formula for the estimated standard error of the mean estimate ($\hat{\sigma}(\hat{X}_M)$). The term in the denominator expressed as $NU_R{}^2S_x{}^2$ is present because the sample size n is present in the finite population correction factor ($\sqrt{1 - n/N}$). Aside from that correction factor, the sample size appears only as the $\sqrt{n}$ factor in the formula for the estimated standard error, and consequently deriving the formula for the sample size requires squaring both sides of the formula for $\hat{\sigma}(\hat{X}_M)$.

A very important feature of the sample size formula is that it expresses the *desired* standard error rather than the *achieved* estimated standard error. Careful comparison of the two formulas will disclose that the desired standard error is expressed by the quantity A/U_R. A denotes the absolute precision desired by the auditor and U_R denotes the reliability factor from the normal distribution corresponding to the interval reliability (R) specified by the auditor.[7]

The derivation of the sample size formula can be illustrated easily by assuming that the finite population correction factor ($\sqrt{1 - n/N}$) has a value of unity (1). With this assumption, the formula for the estimated standard error simplifies to the following:

$$\hat{\sigma}(\hat{X}_M) = \frac{NS_X}{\sqrt{n}}.$$

Squaring both sides of this equation, and solving for n produces this equation:

$$n = \frac{N^2 S_X{}^2}{[\hat{\sigma}(\hat{X}_M)]^2}.$$

Now, by substituting the desired standard error (A/U_R) for $\hat{\sigma}(\hat{X}_M)$, the formula for n becomes this:

$$n = \frac{N^2 S_X{}^2}{(A/U_R)^2} = \frac{N^2 U_R{}^2 S_X{}^2}{A^2}.$$

It is important to note that the auditor's *desired* precision and reliability are explicitly considered in determining the required sample size.

The difficulty in using the formula to determine the sample size is knowing the appropriate value for S_X, the estimated standard deviation of the population of audited amounts. One possibility is to choose a preliminary sample in order to calculate a value for S_X by using the formula given previously. The size of the preliminary sample is important only to the extent that the estimated standard deviation should be reasonably accurate and particularly that it should not be understated. Any percentage error

7. If one-sided reliability (R_1) is desired, the corresponding reliability factor U_{R_1} is used.

in estimating S_X results in about twice that percentage error in the calculated sample size when the finite population correction factor is ignored.[8] Consequently, it is good practice to increase any calculated sample size to allow for possible underestimation of the standard deviation. The practice of adding 10 percent to the calculated sample size protects against the possibility that the standard deviation has been understated by about 5 percent.

The size of any preliminary sample required to provide an estimated standard deviation that is within 5 percent of the actual standard deviation depends upon the distribution of the audited amounts. As long as the final sample size exceeds the preliminary sample size, there is no loss to the auditor in selecting a relatively large preliminary sample.

To provide guidance, the following table of required sample sizes was prepared. In all cases, the desired interval reliability was specified as 95 percent ($U_R = 1.96$). Population sizes (N) are 2,000, 5,000, and 10,000. The desired relative precision (A/X) is either 5 percent or 10 percent.[9] Instead of specifying the value of S_X, the ratio of S_X to $\bar{x}$ is used. This ratio is known as the *coefficient of variation* and for most accounting populations it exceeds one and may be as high as four or five.[10] The table shows this ratio at one, two, three, and four.

Required Sample Size

	Population Size (N)		
	2,000	5,000	10,000
Coefficient of Variation	Relative Precision = .05		
1	869	1176	1332
2	1509	2757	3806
3	1747	3672	5803
4	1850	4155	7109
Coefficient of Variation	Relative Precision = .10		
1	323	357	369
2	869	1176	1332
3	1267	2044	2569
4	1509	2757	3807

This table confirms the fact that large sample sizes are required for unstratified mean estimation. The populations studied by Neter and Loebbecke had coefficients of variation for audited amounts ranging from two to over four. Consequently, preliminary samples as large as 100 would not likely be excessive for the purpose of determining the estimated standard deviation of audited amounts.

8. If S_X increases by an amount ΔS_X so that $\Delta S_X / S_X = t$, then the corresponding increase in sample size is $\Delta n/n = 2t + t^2$. So if $t = .05$, $\Delta n/n = .10 + .0025 = .1025$.
9. Relative precision is expressed as the ratio of the *absolute precision* (A) to the total audited amount (X).
10. To obtain a very rough idea of the coefficient of variation, divide the range of recorded amounts by six times the average recorded amount (total book amount divided by the population size).

When previous statistical experience for a particular population produces a value of the estimated standard deviation, the auditor may use that value adjusted for current conditions rather than employ a preliminary sample. When a preliminary sample is not practical, such as confirmation of accounts receivable, previous nonstatistical experience might be used to obtain a reasonable value for the estimated standard deviation, but this possibility should be only a last resort.

Evaluating Results. Regardless of how the sample size is determined, the results of an unrestricted random sample of n units can be evaluated statistically. The estimated audited amount is the product of the number of units in the population multiplied by the sample mean. In symbols—

$$\hat{X}_M = N\bar{x} = N\frac{\sum x_j}{n}.$$

The achieved precision of the estimate of the total audited amount is calculated using the following formula:

$$A'_M = \frac{NU_R S_X \sqrt{1 - n/N}}{\sqrt{n}}.$$

In this formula, S_X represents the estimated population standard deviation based on the results of the entire sample. The formula may be derived by substituting A'_M/U_R for the estimated standard error of the mean estimator ($\hat{\sigma}(\hat{X}_M)$) in the formula for the estimated standard error given earlier and by then rearranging that formula to solve for A'_M. The achieved precision is denoted by A'_M in order to distinguish it from the desired precision denoted by A.

As explained in chapter 3, the statistical evidence can basically be used either to—

1. Estimate the total audited amount.
2. Decide whether the recorded amount is reasonable.

As also explained there, it may be necessary to adjust the decision interval in case the achieved precision (A'_M) differs from the desired precision (A).

To illustrate these concepts, the following example is presented. The auditor would like to estimate the total value of 6,000 inventory items with a desired absolute precision of $100,000 at 90 percent interval reliability. Since the records contain quantities only, stratification is not practicable. The total recorded amount is $1,120,000 and consequently the auditor expects the average to be about $190. From last year's working papers, the auditor computes an estimate of the standard deviation of $225. Solving for the required sample size, it is found that

$$n = \frac{(6{,}000 \times 1.65 \times 225)^2}{(100{,}000)^2 + 6{,}000 \times (1.65 \times 225)^2}$$
$$= 458.28.$$

Consequently the auditor selects an unrestricted random sample of 459 inventory items, and establishes an audited amount for each selected item. The sample results are as follows:

$\sum x_j = 82{,}620$ (the sum of the sample amounts).

$\sum x_j^2 = 37{,}038{,}800$ (the sum of squares of the sample amounts).

From these results, the auditor computed the estimated total audited amount ($\hat{X}_M$) as

$$\hat{X}_M = 6{,}000 \times \frac{82{,}620}{459} = \$1{,}080{,}000.$$

To compute the precision of this estimate, the auditor first computes the estimated standard deviation of audited amounts as

$$\begin{aligned} S_X &= \sqrt{\frac{37{,}038{,}800 - 459 \times (180)^2}{458}} \\ &= \sqrt{48{,}400} \\ &= 220. \end{aligned}$$

The achieved absolute precision A'_M is then computed as

$$\begin{aligned} A'_M &= \frac{6{,}000 \times 1.65 \times 220\sqrt{1 - \dfrac{459}{6{,}000}}}{\sqrt{459}} \\ &= \frac{2{,}093{,}034}{21.424} \\ &= \$97{,}694. \end{aligned}$$

Since this amount is less than the desired precision of \$100,000, the auditor's objectives are satisfied.

The statistical conclusion is that the recorded amount of \$1,120,000 is overstated by an estimated \$40,000 (\$1,120,000 − \$1,080,000) and with 90 percent reliability, the amount of misstatement ranges from an understatement of \$57,694 (−\$40,000 + \$97,694) to an overstatement of \$137,694 (−\$40,000 − \$97,694).

Difference Estimation

Unstratified difference estimation can be used whenever (1) there is a recorded amount for each sampling unit *and* (2) the distribution of differences is not highly skewed. The existence of a recorded amount for each sampling unit permits defining a difference as the audited amount minus the recorded amount. Whenever the recorded amount exceeds the audited amount (the recorded amount is overstated), the algebraic sign of the difference is negative. Similarly, whenever the recorded amount is less than the audited amount (the recorded amount is understated), the sign of the difference is positive. In other words, the algebraic sign of a difference gives the direction in which the recorded amount should be adjusted to make it agree with the audited amount. A zero difference corresponds to the recorded amount's equalling the audited amount (the recorded amount is correct.)[11]

As a practical matter, unless non-zero differences are reasonably small and both understated differences and overstated differences exist, unstratified difference estimation should probably not be used. The reason for this is that the sample size required may be too large to be practical. Using some form of stratification or pps selection will ordinarily be more efficient.

The difference population will usually have a large proportion of zero values corresponding to correct recorded amounts. This proportion may vary from near 100 percent in the case of well-controlled accounts such as bank demand deposits to less than 50 percent for some priced perpetual inventory records.

The nature of a weakness in internal control may determine the algebraic sign of the non-zero differences. For example, a weakness in recording withdrawals from a material and supply inventory will overstate inventory quantities, a delay in recording sales returns and allowances will overstate accounts receivable, and capitalizing items that should be expensed will overstate fixed assets.

At the present time, little empirical evidence exists concerning the nature, proportion, and amounts of differences in various accounting populations. This makes it impossible to give general guidelines concerning the distribution of differences that may be present.

In spite of little empirical data, two generalizations are possible. First, if differences are all either understatements or overstatements, the difference population will be skewed, but the amount of skewness depends upon both the proportion and magnitudes of the individual non-zero differences. Second, analyzing the nature of the transactions affecting the account and the effectiveness of the controls may enable the auditor to obtain some information about both the proportion of recorded amounts with differences and the potential magnitude of the differences. In this respect, previous experience may also be very helpful.

Sampling Distribution. From an unrestricted random sample of n sampling units, an audited amount is established for each. The individual differences are then computed by subtracting the recorded amounts from the corresponding audited amounts. The estimate of the total difference in the population is equal to the sample average difference multiplied by the total number of sampling units in the population. In symbols, the estimate of the total (or net) difference is

$$\hat{D} = N\bar{d}, \quad \text{where } \bar{d} = \frac{\sum (x_j - y_j)}{n} = \frac{\sum d_j}{n}.$$

The symbol $\bar{d}$ represents the sample mean of the differences whereas d_j represents the difference in the jth sample item. The symbol x_j represents the audited amount of the jth sample item and y_j the recorded amount of that item.

To obtain an estimate of the total audited amount using the difference estimate it is necessary to know the total recorded amount (Y). The estimate of the total audited amount is then equal to the total recorded amount (Y) plus the estimate of the total difference ($\hat{D}$). In symbols,[12]

$$\hat{X}_D = Y + \hat{D}.$$

11. Remember that correctness of the recorded amount pertains only to the auditing procedures employed. Thus a price test of inventory amounts can establish correctness only with respect to prices.
12. The subscript on $\hat{X}_D$ denotes the estimated audited amount is based on the *difference* estimator.

The sampling distribution may be equally well described in terms of either the estimate of the total difference ($\hat{D}$) or the estimate of the total audited amount ($\hat{X}_D$). The mean value of the sampling distribution of the estimate of the total difference is the total population difference. This, of course, is the same as saying that the mean value of the sampling distribution of the estimate of the total audited amount is the total population audited amount. In other words, the difference estimator is unbiased in the same way the mean estimator is.

The standard error of the difference estimator equals the standard deviation of the sampling distribution. This standard deviation is the same whether the sampling distribution refers to the estimate of the total difference or the estimate of the total audited amount. In either case, it is equal to the standard deviation of the population differences divided by the square root of the sample size and multiplied by the product of the population size and the finite population correction factor.

The standard deviation of the population differences is generally not known and must be estimated from the sample results. The estimated standard deviation of the population differences is denoted by the symbol S_D. It can be calculated using the following formula:

$$S_D = \sqrt{\frac{\sum d_j^2 - n\bar{d}^2}{n-1}}.$$

This formula has the same form as the formula for the estimated standard deviation of the population of audited amounts (S_X) except that the sample differences (d_j) are used instead of the sample audited amounts (x_j). Consequently, the operations for computing S_X apply to computing S_D if "difference amounts" is substituted for "audited amounts."

The estimated standard error of the difference estimator is equal to the estimated standard deviation of differences (S_D), divided by the square root of the sample size ($\sqrt{n}$), and multiplied by the product of the population size and the finite population correction factor ($N\sqrt{1 - n/N}$). This estimated standard error is denoted by $\hat{\sigma}(\hat{D})$ [or $\hat{\sigma}(\hat{X}_D)$] and is given by the following formula:

$$\hat{\sigma}(\hat{D}) = \frac{NS_D\sqrt{1 - n/N}}{\sqrt{n}}.$$

Comparing this formula to the estimated standard error of the mean estimator shows that the only change is that the estimated standard deviation of audited amounts (S_X) is replaced by the estimated standard deviation of population differences (S_D).

While the statistical theory says that the difference estimator will be approximately normally distributed for sufficiently large sample sizes, the auditor needs more definite information in order to apply the technique. For sample sizes that are practicable, approximate normality depends upon two factors: (1) the proportion of sampling units with non-zero differences and (2) the distribution of the non-zero differences in terms of both their magnitude and their algebraic sign.

In general, the larger the proportion of sampling units with non-zero differences and the more symmetric the non-zero differences are around zero, the smaller the sample size will need to be for approximate normality to occur. For example, if at least 20 percent of the sampling units have non-zero differences and these differences are nearly equally divided between overstatements and understatements, approximate normality may be achieved with sample sizes in the neighborhood of 100. As the proportion of sampling units with non-zero differences diminishes and/or the distribution of the non-zero differences becomes more skewed, the sample size required will increase. In cases where the proportion of non-zero differences is as small as 5 percent, sample sizes as large as 600 may be necessary for the normal approximation to be valid.[13] Because such large sample sizes are not often practical, the auditor would ordinarily use other techniques such as pps when the proportion of non-zero differences is small.

In addition to approximate normality, the standard deviation of the difference population must be estimated from the sample data. The stability of the estimated standard deviation is affected by the proportion of non-zero differences as well as the distribution of the magnitudes and algebraic sign of the differences. When only a small number of differences is observed, the auditor must be wary that the estimated standard deviation may be well below the actual standard deviation of population differences. An examination of the causes of the differences may help establish the representativeness of the observed differences.

What is the minimum number of differences that must be observed before the standard deviation can be safely estimated? There is no simple answer to this question. A small number would suffice if all the non-zero differences are nearly equal whereas a larger number would be necessary when the differences vary widely. Whatever the number, the auditor should be reasonably satisfied that the observed differences appear to be typical for the particular situation. In that case, recognizing that any numerical guideline has exceptions, the auditor might use 15 or 20 as a minimum number.

Efficiency. The difference estimator is more efficient than the mean estimator whenever the correlation between recorded and audited amounts exceeds one-half. The difference estimator will also be more efficient than the ratio estimator when the magnitudes of the differences are not correlated with the size of the recorded amounts. The ideal situation for the difference estimator would be for each recorded amount to be incorrect by a fairly constant amount rather than by a fairly constant percentage.

Determining Sample Size. Determining a sample size to achieve a desired precision (A) at a specified interval reliability (R) can be done using the following formula:

$$n = \frac{N^2 U_R{}^2 S_D{}^2}{A^2 + N U_R{}^2 S_D{}^2} .$$

This formula has the same form as the corresponding formula for mean estimation. The only difference is that the estimated standard deviation of population differences (S_D) replaces the estimated standard deviation of audited amounts (S_X).

Similar to mean estimation, the difficulty in using this formula is determining a reasonably accurate value for S_D, the estimated standard deviation of the population differences. A preliminary sample offers one method of obtaining information about S_D. If a preliminary sample is used, however, it should be large enough to contain several non-zero differences.

Attribute tables can be used to determine how the sample size is affected by the proportion of non-zero differences in the population because that proportion is equivalent to the population occurrence rate. For example, if the population occurrence

13. These sample sizes were obtained by considering situations where all non-zero differences equal a common value. See Cochran [5], p. 41.

rate is 20 percent, a sample of 90 units will have a 95 percent probability of containing at least 12 observed differences. To contain at least 12 observed differences when the population occurrence rate is only 10 percent, a sample of 180 units is required to achieve the same probability. A sample of 300 has a 95 percent probability of containing at least 12 observed differences when the occurrence rate is 6 percent.

Another method for determining an approximate sample size can be used when all or nearly all of the differences are expected to be overstatements. This method involves obtaining an approximate value for the standard deviation of population differences using the following assumptions:

1. If a recorded amount is incorrect, the amount of overstatement equals the recorded amount.
2. Incorrect accounts are randomly distributed throughout the population.

Since the method calculates an average or expected value for the standard deviation of population differences, it may result in a sample size that is too small to achieve the desired precision.[14] Obtaining an approximation that is too small is not likely, however, when the method is applied to situations where most of the overstatement differences are less than the recorded amounts.

The approximation depends upon the proportion of non-zero differences, the standard deviation of recorded amounts, and the mean of the recorded amounts. To calculate it, the auditor should conservatively estimate the proportion of non-zero differences ($\hat{p}_D$) and perform the following calculations:

1. Multiply the estimated proportion ($\hat{p}_D$) times the square of the standard deviation of recorded amounts (σ_Y^2).
2. Multiply the estimated proportion ($\hat{p}_D$) times its complement ($1 - \hat{p}_D$) and multiply the resulting product times the square of the mean recorded amount ($\bar{Y}^2$).
3. Add the results of the first two steps and take the square root of the sum, $\sqrt{\hat{p}_D\sigma_Y^2 + \hat{p}_D(1 - \hat{p}_D)\bar{Y}^2}$.

Use the resulting number in place of S_D to determine the required sample size.

To illustrate this method, suppose the auditor thought that about 10 percent ($\hat{p}_D = .10$) of 10,000 recorded amounts could contain differences and knew the mean recorded amount ($\bar{Y}$) was \$200 (total recorded amount of \$2 million divided by 10,000) and the standard deviation of recorded amounts (σ_Y) was \$175.[15] Performing the indicated calculations results in the following:

$$\sqrt{.1(\$175)^2 + .1(.9)(\$200)^2} = \$81.62.$$

The rounded amount of \$82 represents an approximation to the standard deviation of population differences that would be expected in these circumstances and may be used in the sample size formula in place of S_D.

Since the auditor seldom knows the exact proportion of non-zero differences, he

14. The technical basis of this method is explained in Appendix 4.
15. When the mean and standard deviation of the recorded population are not known, estimates of these quantities could be used.

should use a conservative estimate in making the calculation. For example, if he felt that the proportion of differences might exceed 10 percent, but would almost certainly not exceed 15 percent, he should set $\hat{p}_D = .15$. Doing so in the above illustration yields a value of $98 instead of $82.

When the value $82 is used in place of S_D in the sample size formula, the resulting sample size is 114 for a desired precision of $50,000 at 95 percent interval reliability. When $98 is used, the resulting sample size increases to 162. Consequently, a 50 percent difference in the proportion of differences leads to about the same difference in the resulting sample size. For comparison, the sample size required for mean estimation would be 497 in order to attain the same level of desired precision ($50,000) at 95 percent interval reliability.[16]

Whatever method is used to compute a sample size, the auditor should consider whether it is large enough to calculate an estimate of the standard deviation of population differences (S_D). In the example, the sample of 114 affords less than an 80 percent chance of including as many as 10 sample differences when 10 percent of the population items have differences. Consequently, the auditor may decide to increase the sample size to, say 180, in order to have an 80 percent chance of observing at least 15 differences or to 220 to have a 95 percent chance of observing at least 15 differences. As an illustration, the sample size is set equal to 180 for this example.

Evaluating Results. The estimate of the total difference ($\hat{D}$) is calculated by multiplying the number of sampling units (N) by the sample average difference ($\bar{d}$). In symbols—

$$\hat{D} = N\bar{d} = N\frac{\sum d_j}{n}.$$

The achieved precision of this estimate is calculated by the following formula:

$$A'_D = \frac{NU_R S_D \sqrt{1 - n/N}}{\sqrt{n}}.$$

The quantity S_D represents the estimated standard deviation of the population differences based on the results of the entire sample.

When the number of observed differences is small, say less than 20, the following procedure may be used to obtain an estimate of the standard deviation of the population differences:

1. From a binomial (or Poisson) table determine the achieved upper precision limit based upon the number of differences observed (m), the sample size (n), and a desired one-sided reliability (R_1). Label this number $P_U(m)$.

2. Calculate the mean difference for only the sample units containing non-zero differences, as

$$\bar{d}_m = \frac{\sum d_j}{m}.$$

16. This was computed assuming the standard deviation of recorded amounts could be used in place of the standard deviation of audited amounts.

3. Calculate the estimated standard deviation for the non-zero population differences as

$$S_D(m) = \sqrt{\frac{\sum d_j^2 - m\bar{d}_m^2}{m - 1}}.$$

4. Determine the estimated standard deviation of the population differences as follows:[17]

$$\hat{\sigma}_D = \sqrt{P_U(m)S_D^2(m) + P_U(m)(1 - P_U(m))(\bar{d}_m)^2}.$$

Finally, calculate the achieved precision (A') using $\hat{\sigma}_D$ in place of S_D and, instead of U_R based on the normal tables, use the corresponding factor from the Student's t-table based on $m - 1$ degrees of freedom.[18]

This suggested procedure increases the estimated standard deviation somewhat to account for the possibility that the occurrence rate of errors in the population could be larger than the fraction of errors observed in the sample. A word of caution is necessary, however. The method does not guard against the possibility that the errors actually observed are unrepresentative of those not observed. The auditor always needs to examine the cause of any observed errors to satisfy himself that any unobserved errors that might exist are likely to be in about the same range of values as those observed in the sample.

For example, suppose that 15 differences were observed in the sample of 180 items previously described. The 15 differences are as follows.

Audited Amt.	Recorded Amt.	Difference	(Difference)^2	(Difference) × (Recorded Amt.)
$ 224	$ 324	$ −100	10,000	−32,400
68	78	− 10	100	− 780
200	299	− 99	9,801	−29,601
300	400	−100	10,000	−40,000
35	65	− 30	900	− 1,950
100	168	− 68	4,624	−11,424
200	242	− 42	1,764	−10,164
79	89	− 10	100	− 890
112	212	−100	10,000	−21,200
67	267	−200	40,000	−53,400
102	192	− 90	8,100	−17,280
61	101	− 40	1,600	− 4,040
175	278	−103	10,609	−28,634
40	42	− 2	4	− 84
50	65	− 15	225	− 975
$1,813	$2,822	$−1,009	$107,827	$−252,822

17. The exact formula is

$$S_D = \sqrt{[(m - 1)/(n - 1)]S_D^2(m) + [m/(n - 1)][1 - (m/n)]\bar{d}_m^2}.$$

18. See Appendix 1 for an explanation of the term *degrees of freedom*.

The mean of the 15 differences is −\$67.27 while the estimated standard deviation for the non-zero population differences is \$53.42. Observing fifteen differences in a sample of 180 yields an achieved upper precision limit for the occurrence rate of differences in the population of .13 at a .95 one-sided reliability. Consequently, following step (4) above produces the following estimated standard deviation of the population differences:

$$\sqrt{.13(53.42)^2 + .13(.87)(67.27)^2} = \$29.71.$$

The t-factor for 14 degrees of freedom is 2.145 at .95 interval reliability. Using these in the formula for the achieved precision gives the following, assuming $N = 10{,}000$:

$$A_D' = \frac{10{,}000 \times 2.145 \times 29.71}{\sqrt{180}} \sqrt{1 - \frac{180}{10{,}000}}$$
$$= \$47{,}071.$$

If, instead of using the approximation, the auditor computes $\bar{d}$ and S_D from the sample data of 180 observations, he finds that $\bar{d} = -\$5.61$ and $S_D = \$23.89$. At a .95 interval reliability using t-factor of 2.145, that yields an achieved precision of

$$A_D' = \frac{10{,}000 \times 2.145 \times 23.89}{\sqrt{180}} \sqrt{1 - \frac{180}{10{,}000}}$$
$$= \$37{,}850.$$

Which of these values is more appropriate to use? The answer depends upon whether the usual estimate ($S_D = \$23.89$) can be regarded as a good estimate of the standard deviation of the difference population. Some empirical study shows that this estimate tends to underestimate the population standard deviation when the number of observed differences is small. As a guideline, it might be reasonable to use the approximation whenever there are fewer than 20 observed differences in the sample and the auditor is reasonably satisfied that the observed differences are representative of the population of differences. This same concern for undervaluing the achieved precision is reflected in the choice of the reliability factor based upon Student's t-table rather than the usual normal table.

For the illustrative data, the estimated total difference is calculated as

$$\hat{D} = 10{,}000 \times (-5.61)$$
$$= \$-56{,}100.$$

Or, equivalently, assuming the recorded amount is $Y = \$2{,}000{,}000$, the estimated audited amount is

$$\hat{X}_D = \$2{,}000{,}000 - 56{,}100$$
$$= \$1{,}943{,}900.$$

Using this procedure for obtaining a value for the achieved precision of the difference estimate requires observing some differences in the sample. What can be done if no differences are observed or if only a few differences are observed and the auditor cannot ascertain that these are reasonably representative? There are several possibilities.

The auditor can increase the sample. This is appropriate when only a few differences are observed and their representativeness cannot be established. If the sample size cannot be increased or if no differences are observed, the auditor might consider one of the following.

A conservative bound for overstatement error may be determined by multiplying the largest recorded amount times the achieved upper precision limit for the occurrence rate of errors in the population. The achieved upper precision limit may be determined at a desired one-sided reliability (R_1). This method has the advantage of not depending on the approximate normality of the sampling distribution, but the disadvantage that the bound produced is frequently very large.

Two other methods that do depend on the approximate normality of the sampling distribution are the following. Use the standard deviation of the recorded amounts in place of S_D or use the combined mean-per-unit and difference estimator described by Neter and Loebbecke [18]. If the latter method is used, the weight (w) should be selected before the sample items are audited.

To illustrate these methods, suppose that the sample of 180 items previously described resulted in no sample differences. Suppose further that the largest recorded amount is $1,100. Observing zero differences in a sample of 180 yields an upper precision limit of .02 at a one-sided reliability of .95. A conservative bound for the overstatement error is $220,000 ($1,100 × .02 × 10,000) at a one-sided reliability of .95.

Using the standard deviation of recorded amounts ($175) in place of S_D in the formula for the achieved precision results in a value of about $211,000 at a one-sided reliability ($U_{R_1} = 1.65$) of .95:

$$\$211{,}348 = \frac{10{,}000 \times 1.65 \times \$175}{\sqrt{180}} \sqrt{1 - \frac{180}{10{,}000}}.$$

Note that both of these bounds exceed the desired precision of $50,000. This is to be expected. The sample size (180) in this example was determined on the basis that about 10 percent of the population items were in error. Had the auditor anticipated observing no errors, he should have planned the sample size so that the resulting bound on overstatement error would be the desired size.

Ratio Estimation

Unstratified ratio estimation can be used whenever (1) there is a positive recorded amount for each sampling unit, *and* (2) the distribution of ratios is not highly skewed.

The population ratio is defined as the total audited amount divided by the total recorded amount. Whenever this ratio is less than one (the total audited amount is less than the total recorded amount) the population recorded amount is overstated. If the ratio is larger than one (the total audited amount exceeds the total recorded amount) the population recorded amount is understated. A necessary assumption is that the population recorded amounts are all positive.

Sampling Distribution. From an unrestricted random sample of n sampling units, the auditor establishes the audited amounts together with the corresponding recorded amounts. The auditor can then form the estimated population ratio by dividing the total sample audited amount by the total sample recorded amount. This is represented in symbols thus:

$$\hat{R} = \frac{\sum x_j}{\sum y_j}.$$

The estimate of the total audited amount is then computed by multiplying the estimated population ratio times the known recorded amount. Thus,[19]

$$\hat{X}_R = \hat{R}Y.$$

Alternatively, the auditor may base the estimated population ratio on the observed sample differences (audited amounts minus recorded amounts). In this case, the estimated population ratio equals one (1.0) plus the total sample difference divided by the total sample recorded amount. This is expressed symbolically as follows:

$$\hat{R} = 1 + \frac{\sum d_j}{\sum y_j}.$$

This permits estimating the total difference amount (total audited amount minus total recorded amount) by multiplying the total recorded amount by the ratio of the total sample difference divided by the total sample recorded amount. That is,

$$\hat{D}_R = \frac{\sum d_j}{\sum y_j} Y = (\hat{R} - 1)Y.$$

Averaged over all possible samples the mean value of $\hat{X}_R$ does not equal the population audited amount, but the difference is small when the sample is large. In other words, the ratio estimate of the total audited amount is biased, but the bias is negligible for large samples.

The standard error of the ratio estimate of the total audited amount is equal to the standard deviation of the sampling distribution. This standard deviation is equal to the standard deviation of the population ratios divided by the square root of the sample size and multiplied by the product of the population size and the finite population correction factor.

The standard deviation of the population ratios is not known in general and consequently must be estimated. The estimated standard deviation of population ratios is denoted by the symbol S_R and may be calculated from the following formula:

$$S_R = \sqrt{\frac{\sum x_j^2 + \hat{R}^2 \sum y_j^2 - 2\hat{R} \sum x_j y_j}{n - 1}}.$$

An equivalent formula that depends on the differences is this:

$$S_R = \sqrt{\frac{\sum d_j^2 + (\hat{R} - 1)^2 \sum y_j^2 - 2(\hat{R} - 1) \sum d_j y_j}{n - 1}}.$$

19. The subscript R on $\hat{X}_R$ indicates that the estimated total audited amount is based on the *ratio* estimator.

The estimated standard error of the ratio estimate of the total audited amount is equal to the estimated standard deviation of population ratios (S_R) divided by the square root of the sample size ($\sqrt{n}$) and multiplied by the product of the population size and the finite correction factor ($N\sqrt{1 - n/N}$). This estimated standard error is denoted by $\hat{\sigma}(\hat{X}_R)$ and is determined by the following formula:

$$\hat{\sigma}(\hat{X}_R) = \frac{NS_R\sqrt{1 - n/N}}{\sqrt{n}} .$$

Cochran [5] suggests that the normal approximation will be valid provided the sample size is large enough to make both the coefficient of variation of recorded amounts and the coefficient of variation of audited amounts less than 10 percent when divided by the square root of the sample size.[20] The coefficient of variation of recorded amounts can be easily computed when the standard deviation of recorded amounts is known—otherwise it can be estimated from a sample of the recorded amounts. The coefficient of variation of audited amounts is not known, but, for planning purposes the auditor may often assume that it does not differ much from the coefficient of variation of recorded amounts.

One method of accounting for the possibility that the coefficient of variation of audited amounts might be larger than the coefficient of variation of recorded amounts is to divide the coefficient of variation of recorded amounts by the factor ($1 - M/Y$), where M represents the amount considered material and Y is the total recorded amount. For example, if M = \$100,000 while Y = \$1 million, the factor equals .9 (1 – (100,000/1,000,000)), and the coefficient of variation of audited amounts equals 1.11 times the coefficient of variation of recorded amounts.[21]

To achieve the quotient of 10 percent for the recorded amounts, the sample size must be at least 100 times the square of the coefficient of variation of recorded amounts. Since the coefficient of variation may range from 1 to 5 for many accounting populations, the minimum sample sizes typically will range from 100 to 2,500 items. The minimum sample size should be increased for the possibility that the coefficient of variation of audited amounts is larger than the coefficient of variation of recorded amounts. In many cases a safe amount of increase would be dividing by the factor $(1 - M/Y)^2$ corresponding to the above cited relationship. For example, if $1 - M/Y = .9$, the sample size should be increased by 23.5 percent ($1/.9^2 = 1/.81 = 1.235$).

In addition to Cochran's suggestion, the sampling distribution is affected by the extent and magnitudes of the differences between audited and recorded amounts. When the proportion of non-zero differences in the population is very small (less than 5 percent, say) many samples contain no differences at all. This precludes making reasonably accurate estimates of the standard deviation of the ratio population. Just as in the case of difference estimation, it is not possible to give an ironclad rule on

---

20. The coefficient of variation equals the standard deviation divided by the mean.
21. The rationale for using this factor is that if the audited amount is less than the recorded amount by a material amount, while the standard deviation of audited amounts is about the same as the standard deviation of recorded amounts,

$$\frac{\sigma_X}{\bar{X}} = \frac{\sigma_Y}{\bar{Y} - \frac{M}{N}} = \frac{\sigma_Y}{\bar{Y}}\left(\frac{Y}{Y - M}\right) = \frac{\sigma_Y}{\bar{Y}}\left(\frac{1}{1 - \frac{M}{Y}}\right) .$$

the minimum number of non-zero differences required, and the suggestions made there apply equally well to ratio estimation.

The distribution of the differences affects both the validity of the normal approximation and the efficiency of the ratio estimator. The papers of Kaplan [14] and Neter and Loebbecke [18] suggest that when differences are all of one sign—either all understatements or all overstatements—the sampling distribution can fail to be approximately normal even in the presence of many differences. Further empirical study is necessary to clarify whether the sign of the difference is responsible for this or whether there is some other cause.

Efficiency. The ratio estimator is most efficient when each audited amount is nearly proportional to the recorded amount. The ideal situation for the ratio estimator is for each audited amount to be equal to a constant multiple of the recorded amount. In this ideal situation, the difference between each audited amount and recorded amount would also be a constant multiple of the recorded amount.

The ratio estimator is more efficient than the mean estimator whenever the correlation between recorded and audited amounts exceeds one-half.[22] The ratio estimator will also be more efficient than the difference estimate if the magnitude of the differences is highly correlated with the recorded amounts. Finally, the ratio estimator will be about as efficient as the regression estimator when the ideal situation is approximately satisfied (the regression line passes through the origin).

Determining Sample Size. Sample size is determined to achieve a desired precision at a specified reliability. The sample should be large enough to assure the validity of the normal approximation and afford the opportunity to observe several differences. The following formula may be used in determining the appropriate sample size:

$$n = \frac{N^2 U_R^2 S_R^2}{A^2 + N U_R^2 S_R^2}.$$

Again, this formula is quite similar to the corresponding formula for both difference estimation and mean estimation. The only distinction is that the estimated standard deviation for each technique is calculated in a different way.

The following table of required sample sizes for unstratified ratio estimation was constructed to be comparable to a similar table (see page 69) for mean estimation. For both tables, the desired interval reliability (R) is .95 ($U_R = 1.96$), and the desired relative precision (A/X) is either 5 percent or 10 percent. The coefficient of variation for audited amounts is taken to be one, two, three, or four in both tables. Additionally, the following table considers the coefficient of variation of recorded amounts as being approximately equal to the coefficient of variation of audited amounts. To demonstrate the effect of correlation between audited and recorded amounts, the table below reflects correlation coefficients at .50, .80, and .95. The population size is taken to be 5,000 (the mean estimation table additionally covers population sizes of 2,000 and 10,000).

22. This assumes that the coefficient of variation of audited amounts is at least as large as the coefficient of variation of recorded amounts.

Required Sample Size			
	Population Size (N) = 5,000 Correlation Coefficient		
	.50	.80	.95
Coefficient of Variation	Relative Precision = .05		
1	1176	548	149
2	2757	1648	548
3	3672	2626	1084
4	4155	3315	1649
Coefficient of Variation	Relative Precision = .10		
1	357	149	38
2	1176	548	149
3	2044	1083	324
4	2757	1648	548

In the above table the first column that corresponds to a correlation coefficient of .50 shows the same sample size that is required when mean estimation is used. As the correlation between recorded and audited amounts rises, the reduction in the sample size required when using the ratio estimator is demonstrated by comparing the second and third columns with the first column in the foregoing table.

To use the formula for determining the sample size, a reasonably accurate estimate of S_R is required. A preliminary sample may be used to estimate S_R, but the sample must be large enough to contain several non-zero differences. Attribute tables can be used to determine the sample size required to exhibit a specified number of sample differences. Use of a binomial table for guidance in this respect was discussed in determining the sample size for difference estimation.

When all or nearly all of the differences are expected to be overstatements, the auditor can obtain an approximation for the required sample size by assuming that any overstatement error is equal to the recorded amount and that incorrect accounts are randomly distributed throughout the population. The approximation entails calculating an average or expected value of the standard deviation of population ratios to be used in place of S_R.[23]

As in the case of difference estimation, the approximation depends upon the proportion of non-zero differences, the standard deviation of recorded amounts (σ_Y) and the mean of the recorded amounts ($\bar{Y}$). The auditor should conservatively estimate the proportion of non-zero differences ($\hat{p}_D$) and perform the following calculations:

1. Multiply the estimated proportion ($\hat{p}_D$) times its complement ($1 - \hat{p}_D$).
2. Multiply this product by the *sum* of the square of the standard deviation of recorded amounts (σ_Y^2) and the square of the mean recorded amount ($\bar{Y}^2$).
3. Take the square root of the resulting number, $\sqrt{\hat{p}_D(1 - \hat{p}_D)(\sigma_Y^2 + \bar{Y}^2)}$.

23. The technical basis of this method is explained in Appendix 4.

Use the resulting number in place of S_R to determine the required sample size.

The resulting approximation is always smaller than the corresponding approximation for the difference estimator. For example, using the same data as in the case of difference estimation ($\hat{p}_D = .10$, $\sigma_Y = \$175$, $\bar{Y} = \$200$) produces the following result:

$$\sqrt{(.10)(.90)((175)^2 + (200)^2)} = \$79.73.$$

Using the rounded amount of \$80 in place of S_R in the formula for determining sample size yields a required sample size of 108 for the desired precision of \$150,000 at 95 percent interval reliability. This sample size, however, has only about a 13 percent chance of including as many as 15 sample differences when 10 percent of the population items have differences. Consequently, just as in the case of difference estimation, the auditor may decide to increase the sample size to 180 in order to have an 80 percent chance of observing at least 15 sample differences, or to 220 to increase the chance to 95 percent.

Evaluating Results. The estimated audited amount using ratio estimation is

$$\hat{X}_R = \hat{R}Y,$$

where $\hat{R}$ is the estimated ratio. As previously explained, the estimated ratio $\hat{R}$ may be calculated either as the ratio of the sum of the sample audited amounts to the sum of the sample recorded amounts or as 1.0 plus the ratio of the sum of the sample difference amounts to the sum of the sample recorded amounts.

The achieved precision of this estimate is calculated by the following formula:

$$A'_R = \frac{NU_R S_R \sqrt{1 - n/N}}{\sqrt{n}},$$

where S_R is the estimated standard deviation of the population ratios. Two formulas for S_R were previously presented, one using the sample audited amounts (x_j) and the other using the sample difference amounts ($d_j = x_j - y_j$). The one using the differences is generally simpler to use because sample differences will be zero when the recorded amounts and audited amounts agree.

When few differences are observed in the sample, the estimated standard deviation (S_R) may underestimate the population standard deviation. The following procedure may be helpful in obtaining an estimate of the standard deviation of population ratios, provided the differences that are observed appear representative.

1. From a binomial (or Poisson) table determine the achieved upper precision limit for the proportion of differences based upon the observed number of differences (m), the sample size (n), and a desired one-sided reliability (R_1). Label this number $P_U(m)$.

2. Calculate $S_R(m)$ in the following way:

   $$S_R^2(m) = \frac{n - 1}{m - 1} S_R^2 .$$

3. Determine the estimated standard deviation of population ratios as

   $$\hat{\sigma}_R = \sqrt{P_U(m) S_R^2(m)}.$$

In addition, when computing the achieved precision (A'_R), instead of using U_R based on the normal tables, use the corresponding factor from the Student's t-table with $(m - 1)$ degrees of freedom.

This suggestion, just as the one made for difference estimation, only compensates for the possibility that the proportion of differences in the sample is smaller than the proportion of differences in the population. As before, the auditor needs to examine the causes of any observed errors before proceeding with the statistical analysis.

When either no differences are observed or there is considerable doubt concerning the representativeness of those observed, the auditor might consider one of the bounds on overstatement error discussed in the difference estimation section. Lack of representativeness may also cause the auditor to increase the sample size.

To illustrate, consider the sample of 180 items with 15 observed differences as shown in the section dealing with difference estimation. In addition to the data shown there, suppose that the sum of recorded amounts over the whole sample is

$$\sum y_j = 37{,}620, \text{ so that } \hat{R} - 1 = \frac{-1{,}009}{37{,}620} = -.027,$$

and the sum of squares of recorded amounts is

$$\sum y_j^2 = 12{,}444{,}980.$$

A straightforward application of the formula for S_R gives

$$S_R = \sqrt{\frac{107{,}827 + (.027)^2(12{,}444{,}980) - 2(-.027)(-252{,}822)}{179}}$$

$$= \$24.00.$$

Applying the suggested procedure, the value of $P_U(m)$ is .13 as before, and

$$S_R^2(m) = \frac{179}{14}(576)$$

$$= 7364.57.$$

Consequently,

$$\hat{\sigma}_R = \sqrt{.13(7364.57)}$$

$$= \$30.94 \quad (\$31.00 \text{ rounded}).$$

As before, the t-factor for 14 degrees of freedom is 2.145 at .95 interval reliability. Consequently, the achieved precision is given by

$$A'_R = \frac{10{,}000 \times 2.145 \times 31.00\sqrt{1 - \dfrac{180}{10{,}000}}}{\sqrt{180}}$$

$$= \frac{652{,}981}{13.416}$$

$$= \$48{,}672.$$

If, instead, the auditor uses the computed value of S_R($24.00), the achieved precision is calculated as

$$A'_R = \frac{10{,}000 \times 2.145 \times 24.00\sqrt{1 - \dfrac{180}{10{,}000}}}{\sqrt{180}}$$

$$= \frac{505{,}534}{13.416}$$

$$= \$37{,}681.$$

The larger value provides some protection against the possibility that the 15 observed differences give an estimate that is too small because the proportion of differences in the population exceeds the proportion of differences in the sample. As in the case of differences, no definite rule can be given concerning when to use the approximate value, but, as a guideline, the auditor might consider doing so whenever the number of observed differences is fewer than 20. However, the auditor should be reasonably satisfied that the observed differences are representative.

For this data, the estimated ratio is

$$\hat{R} = .973,$$

and, assuming that the total recorded amount is $2 million, the estimated audited amount is

$$\hat{X}_R = (.973)(\$2{,}000{,}000) = \$1{,}946{,}000;$$

or, equivalently, the estimated total difference is

$$\hat{D}_R = (-.027)(\$2{,}000{,}000) = -\$54{,}000.$$

These estimates are close to the difference estimates for this illustration. Such close agreement is typical of most applications.

Regression Estimation

When recorded amounts are available and their distribution is not highly skewed, unstratified regression estimation can be used instead of either difference estimation or ratio estimation. The regression estimator of the total audited amount is given by the following formula:[24]

$$\hat{X}_G = N\bar{x} + b(Y - N\bar{y}),$$

where $\bar{x}$ is the sample mean audited amount, b is the estimated regression coefficient, Y is the total recorded amount, and $\bar{y}$ is the sample mean recorded amount. When $b = 0$, $\hat{X}_G$ equals the mean estimator of the total audited amount, when $b = 1$, it equals the difference estimator of the total audited amount, and when $b = \bar{x}/\bar{y}$, it equals the ratio estimator of the total audited amount.

24. The subscript G on $\hat{X}_G$ indicates that the estimated total amount is based on the *regression* estimator.

The value of b in the regression estimate of the total audited amount may be determined using the following formula:

$$b = \frac{\sum x_j y_j - n\bar{x}\bar{y}}{\sum y_j^2 - n\bar{y}^2}.$$

Alternatively, b may be determined by using a formula based on the observed differences ($d_j = x_j - y_j$) as follows:

$$b = 1 + \frac{\sum d_j y_j - n\bar{d}\bar{y}}{\sum y_j^2 - n\bar{y}^2}.$$

When only some of the sample items have non-zero differences, this latter formula is easier to compute.

Using this latter formula for b, it is possible to express the regression estimate of the total audited amount as follows:

$$\hat{X}_G = Y + N\bar{d} + \frac{\sum d_j y_j - n\bar{d}\bar{y}}{\sum y_j^2 - n\bar{y}^2}(Y - N\bar{y}).$$

Subtracting the total recorded amount (Y) from both sides of this equation yields a regression estimate of the total difference between audited and recorded amounts ($\hat{D}_G$) as follows:

$$\hat{D}_G = N\bar{d} + (b - 1)(Y - N\bar{y}).$$

Sampling Distribution. Over all possible samples the mean value of $\hat{X}_G$ does not equal the total audited amount, but the difference is small for large sample sizes. In other words, the regression estimator is biased, but the bias is negligible for large sample sizes.

The estimated standard error of the regression estimate of the total audited amount is given by the following formula:

$$\hat{\sigma}(\hat{X}_G) = \frac{NS_G\sqrt{1 - n/N}}{\sqrt{n}},$$

where S_G is the estimated standard deviation of the regression population.[25] The estimated standard deviation may be calculated from the following formula:

$$S_G = \sqrt{\frac{1}{n-2}\left[\sum x_j^2 - n\bar{x}^2 - \frac{(\sum x_j y_j - n\bar{x}\bar{y})^2}{\sum y_j^2 - n\bar{y}^2}\right]}.$$

25. The term *regression population* is used here to denote the population residuals where the residuals refer to the differences between the audited amounts and the "true" regression.

An equivalent formula that uses the differences instead of the audited amounts is given by the following:

$$S_G = \sqrt{\frac{1}{n-2}\left[\sum d_j^2 - n\bar{d}^2 - \frac{(\sum d_j y_j - n\bar{d}\bar{y})^2}{\sum y_j^2 - n\bar{y}^2}\right]}\ .$$

The approximate normality of the sampling distribution of the regression estimator is affected by (1) the distribution of the recorded amounts, (2) the proportion of non-zero differences, and (3) the distribution of the non-zero differences.

The situation parallels that of the ratio estimator and similar conclusions are relevant. In particular, the coefficient of variation of the recorded amounts divided by the square root of the sample size should be less than 10 percent, and the sample should contain several non-zero differences. It is not possible to give a definite number, but, as before, using 15 or 20 as a minimum number when the observed differences appear typical for the particular situation seems to be a reasonable operating rule.

Efficiency. The regression estimator is more efficient than the mean estimator whenever the correlation between the audited amounts and recorded amounts is different from zero. This will nearly always be the case.

The regression estimator is likewise more efficient than the difference estimator unless the magnitudes of the differences are unrelated to the size of the recorded amounts. Even in this unlikely situation, the regression estimator is as efficient as the difference estimator.

The regression estimator is also just as efficient as the ratio estimator in all cases, and more efficient in most. The ratio estimator is nearly as efficient only when each audited amount is nearly proportional to the corresponding recorded amount. For populations containing a large proportion of recorded amounts that equal the audited amounts, the condition of near proportionality is satisfied only when the non-zero differences are small in magnitude.

The conclusion is that the regression estimator is the most efficient estimator of the four. Furthermore, it does not require that all recorded amounts be positive as does the ratio estimator.

Determining Sampling Size. To determine the sample size required to achieve a desired precision (A) at a specified interval reliability (R), the following formula may be used:

$$n = \frac{N^2 U_R^2 S_G^2}{A^2 + N U_R^2 S_G^2}\ .$$

This formula uses the estimated standard deviation of the regression population in the same way that the other estimation techniques employed the required estimated standard deviation.

Similar to the preceding table for ratio estimation, the following table shows the required sample sizes for regression estimation when the interval reliability (R) is .95 ($U_R = 1.96$); the population size is 5,000; the desired relative precision (A/X) is either 5 percent or 10 percent; the coefficient of variation of audited amounts is one, two, three, or four and equals the coefficient of variation of recorded amounts; and the correlation coefficient is .50, .85, or .95.

Required Sample Size

	Population Size (N) = 5,000 Correlation Coefficient		
	.50	.80	.95
Coefficient of Variation	Relative Precision = .05		
1	936	498	146
2	2398	1534	535
3	3374	2495	1062
4	3933	3195	1620
Coefficient of Variation	Relative Precision = .10		
1	272	134	37
2	937	498	146
3	1708	997	316
4	2399	1534	535

Comparing the foregoing table to the similar table concerning ratio estimation demonstrates that, in identical circumstances, regression estimation requires smaller sample sizes than ratio estimation. However, comparing the right-hand columns in the two tables shows that the advantage of regression estimation is very small when the correlation coefficient is .95.

The left-hand column of the ratio estimation table, corresponding to a correlation coefficient of .50, also gives the sample sizes required for mean estimation of audited amounts. Comparing the left-hand columns of the two tables also shows that regression estimation requires smaller sample sizes than mean estimation. The advantage of regression estimation over mean estimation persists whenever the correlation coefficient is different from zero.

The above formula for determining an appropriate sample size requires the auditor to have a reasonably accurate estimate of S_G. If a preliminary sample is used to estimate S_G, it should be large enough to contain several non-zero differences. As with the other estimation techniques, attribute tables can be used to determine a sample size necessary to contain a specified number of sample differences.

When all or nearly all of the differences are expected to be overstatements, the auditor can determine an approximation to the required sample size by assuming that any overstatement error is equal to the recorded amount. As was described for difference and ratio estimation, the approximation entails calculating an average or expected value for the standard deviation of the regression population to be used in place of S_G.

The approximation in the regression case is exactly the same as described for the ratio case. This equality occurs because under the assumed form of the differences, the population regression and the population ratio exactly coincide.[26] The illustration presented for ratio estimation applies also to regression estimation, including

26. Another way of expressing this is that the regression line passes through the origin.

the suggestion to increase the sample size in order to assure observing several differences.

Evaluating Results. The estimated audited amount using regression estimation is

$$\hat{X}_G = N\bar{x} + b(Y - N\bar{y}),$$

where b is the estimated regression coefficient based on the entire sample.

The achieved precision of the regression estimate of the total audited amount is calculated as follows:

$$A'_G = \frac{NU_R S_G}{\sqrt{n}} \sqrt{1 - n/N},$$

where S_G is the estimated standard deviation of the regression population.

Just as for difference and ratio estimation, the estimated standard deviation (S_G) may underestimate the population standard deviation when only a few differences are observed. The same suggestions made in the section on difference estimation apply in case the auditor observes no differences or does not think the observed differences are representative. The following procedure may be used to obtain an estimate of the standard deviation of the regression population in cases where the auditor feels the proportion of differences in the sample may be smaller than the proportion of differences in the population.

1. From a binomial (or Poisson) table determine the achieved upper precision limit for the proportion of differences based upon the observed number of differences (m), the sample size (n), and a desired one-sided reliability (R_1). Label this number $P_U(m)$.

2. Calculate the mean difference for only the sample units containing non-zero differences, as follows:

$$\bar{d}_m = \frac{\sum d_j}{m}.$$

3. Calculate the estimated standard deviation for the non-zero population difference as follows:

$$S_D(m) = \sqrt{\frac{\sum d_j^2 - m\bar{d}_m^2}{m - 1}}.$$

4. Calculate the quantity

$$\frac{1}{n-2} \frac{(\sum d_j y_j - m\bar{d}_m \bar{y}_n)^2}{\sum y_j^2 - n\bar{y}^2} = \left(\frac{n-1}{n-2}\right)(b - 1)^2 S_Y^2,$$

where $\bar{y}_n$ is the average recorded amount for the entire sample.

5. Determine the estimated standard deviation of the regression population as follows:

$$\hat{\sigma}_G = \sqrt{P_U(m)S_D^2(m) + P_U(m)(1 - P_U(m))\bar{d}_m^{\,2} - \left(\frac{n-1}{n-2}\right)(b-1)^2 S_Y^{\,2}}.$$

In addition, when computing the achieved precision (A_G'), use the reliability factor from the Student's t-table with $(m - 2)$ degrees of freedom instead of the U_R based on the normal tables.

To illustrate, the same sample of 180 items with 15 observed differences that was previously evaluated by difference and ratio methods is evaluated using the regression technique. Using the formula for S_G with $\sum d_j^2 = 107{,}827$, $\sum d_j = -1{,}009$, $\sum d_j y_j = (-252{,}822)$, $\bar{y} = 209$, $\sum y_j^2 = 12{,}444{,}980$ and $n = 180$, produces the following result

$$\begin{aligned} S_G &= \sqrt{\frac{1}{178}\left[107{,}827 - 5{,}656 - \frac{(-252{,}822 - 210{,}881)^2}{12{,}444{,}980 - 7{,}862{,}580}\right]} \\ &= \sqrt{\frac{101{,}787}{178}} \\ &= \$23.91. \end{aligned}$$

Applying the suggested procedure, with $P_U(m) = .13$ as before gives

$$\begin{aligned} \hat{\sigma}_G &= \sqrt{.13(53.42)^2 + .13(.87)(67.27)^2 - \frac{179}{178}(.0092)^2(160)^2} \\ &= \sqrt{370.98 + 511.81 - 2.18} \\ &= \$29.68. \end{aligned}$$

In this latter calculation, it was necessary to compute the value of b from the formula as

$$\begin{aligned} b &= 1 + \frac{-252{,}822 + 210{,}881}{12{,}444{,}980 - 7{,}862{,}580}, \\ &= .9908 \end{aligned}$$

and $S_Y = \sqrt{\dfrac{12{,}444{,}980 - 7{,}862{,}580}{179}} = 160.$

As before, the auditor should consider using the approximation $\hat{\sigma}_G$ instead of S_G whenever the observed number of sample differences is small—say, fewer than 20—and his investigation of the observed errors supports their being representative. In this illustration, the computed value of $\hat{\sigma}_G$(\$29.68) is very close to the previously computed value of $\hat{\sigma}_D$ (\$29.71), reflecting the fact that b is very close to one and hence the regression estimate, in this case, is quite close to the difference estimate.

To compute the value of the regression estimate, the formula gives,

$$\begin{aligned} \hat{X}_G &= (10{,}000)(\$203.39) + (.9902)(\$2{,}000{,}000 - (10{,}000)(\$209)) \\ &= \$2{,}033{,}900 + .9902(-\$90{,}000) \\ &= \$1{,}944{,}782 \quad (\$1{,}944{,}800 \text{ rounded}), \end{aligned}$$

or equivalently, the regression estimate of the difference is[27]

$$\hat{D}_G = \$2{,}000{,}000 - \$1{,}944{,}782$$
$$= \$55{,}218.$$

The achieved precision of either of these estimates may be computed using the formula on page 90 with 2.145 being used as the reliability factor. If $\hat{\sigma}_G$ is used in place of S_G, the precision is computed as

$$A'_G = \frac{10{,}000 \times 2.145 \times \$29.68 \times \sqrt{1 - \dfrac{180}{10{,}000}}}{\sqrt{180}}$$
$$= \frac{630{,}880}{13.416}$$
$$= \$47{,}023.$$

Using S_G, the precision may be computed as

$$A'_G = \frac{10{,}000 \times 2.145 \times \$23.91 \times \sqrt{1 - \dfrac{180}{10{,}000}}}{\sqrt{180}}$$
$$= \$37{,}883.$$

Summary

Four variable sampling methods—mean, difference, ratio, and regression estimation—have been discussed in the case of unstratified sampling. Unstratified sampling has limited usefulness in auditing even when the largest sampling units are selected on a 100 percent basis. The basic concepts discussed in this chapter can be extended to the more practical case of stratified sampling, which is discussed in the next chapter.

When recorded amounts exist for each of the sampling units, and their total agrees with the general-ledger control amount, the auditor can select a difference, ratio, or regression estimator of the total audited amount. If at most a few differences are expected, the auditor can plan the sample size to yield a useful upper bound on the overstatement error. This can be done either by using the attribute tables (or computer program) or by using the sample size formula corresponding to the mean estimator.

When several differences are expected, the auditor can use one of the formulas corresponding to the difference, ratio, or regression estimator supplemented by the attribute tables (or computer program). This gives some assurance of observing enough differences to make valid inferences about the extent of difference in the population. For each of the three estimators, an approximation may be useful in obtaining a preliminary estimate of the appropriate standard deviation in cases where the auditor anticipates mostly overstatement errors.

---

27. The value of \$203.39 for $\bar{x}$ can be calculated from the data given since $\Sigma x_j = \Sigma y_j + \Sigma_m x_j - \Sigma_m y_j = 37{,}620 + 1813 - 2822 = 36{,}611$.

After obtaining an audited amount for each of the sample items, the auditor can compute an estimate of the total audited amount based on either the difference, ratio, or regression formula. When many differences are present, say more than 20, the auditor can use the appropriate formula for computing the achieved precision of the estimate. If the number of differences observed is between, say, 10 and 20, and the auditor is reasonably satisfied that the observed differences are representative of the population of differences, an approximation can be used to compute the achieved precision. If the number of differences is small or the auditor is not satisfied that the observed differences are representative, either one of the bounds for overstatement can be used or the sample can be enlarged.

In many circumstances, the three estimators give nearly the same results although the regression estimator is the most efficient. For the data evaluated in the chapter, the following table shows the comparative results.

	Difference	Ratio	Regression
Estimated audited amount	$1,943,900	$1,946,000	$1,944,800
Recorded amount	2,000,000	2,000,000	2,000,000
Estimated difference	−56,100	−54,000	−55,000
Achieved precision at .95 interval reliability	37,850	37,681	37,883
Approximate precision at .95 interval reliability	47,071	48,672	47,023

6

Stratified Variable Methods

Currently, there are two methods widely used in practice to select variable samples—stratified random sampling and pps sampling. Each method provides greater representation of the larger recorded amounts in the sample in contrast to unrestricted random sampling which provides equal representation of every recorded amount. Under both methods, the required sample sizes will generally be much less than the size required when using unrestricted random sampling.

The auditor would normally select one of these methods whenever planning a statistical substantive test of an account balance or class of transactions. In current practice, inventories and accounts receivable comprise the major areas of application. In nearly all cases, a computer is used to assist the auditor.

Both methods are discussed in this chapter, and formulas—often complex—are used throughout, illustrated, in most cases, with numerical examples. However, the computer programs described in chapter 9 make it unnecessary for the auditor to use these formulas for making calculations.

The purpose of stratification is to improve efficiency by reducing sample sizes for desired levels of precision and reliability. The previous illustrations of the required sample sizes using unstratified random sampling demonstrate how very large the sample must be when the population variability is high (measured by the coefficient of variation). Such large sample sizes are not practical in auditing—stratification is one technique that will allow reasonable sample sizes.

Stratified random samples achieve efficiency by grouping sampling units with similar characteristics into separate strata. In this way the variability among sampling units within any one stratum is small. Ideally, the strata would be defined so that all sampling units within a stratum had the same value—in that case a sample consisting of one unit from each stratum would constitute the most efficient sample possible. This ideal cannot be achieved for two reasons: (1) It would require too many strata to be practical, and (2) it is necessary to classify the sampling units into strata prior to sampling.

The first reason merely states the obvious—to achieve the ideal would require as many strata as there are different values in the population, and this would ordinarily be an impractically large number.

The second reason is more important. It indicates that stratification must be based on information available to the auditor before any auditing can occur. For example, if the sampling units are priced perpetual inventory records, the basis for stratifying the sampling units must be among the items of information on the record, such as recorded amounts, type of item, storage location, or volume of activity. In choosing among the possibilities, the auditor considers the following:

1. What basis will be most efficient?

2. What basis will be least expensive to implement?

The comparative efficiency of different stratification bases is determined by the relationship between the basis and the particular standard deviation involved. For example, with mean estimation of the total audited amount, the auditor would like to choose a basis that would result in sampling units with similar audited amounts being in a stratum. Likewise, with difference estimation of the total audited amount the auditor would prefer a basis that would place sampling units with similar differences in a stratum. In the inventory example, the recorded amount might be the best choice as a stratification basis for mean estimation, while the volume of activity might be the best choice for difference, ratio, or regression estimation.

Cost is another important factor that must be considered. Manual stratification of a large file is both time-consuming and expensive. Consequently, the only method of stratifying a population that appears practical for auditing is using a computer. Most computer programs, including the ones designed to accompany this book, use a quantitative field as the basis of stratification. This restricts the choice to a quantity —such as a recorded amount—that is available in the record.

Stratification by Recorded Amounts

Recorded amounts are currently widely used as the basis for stratifying the population. While recorded amounts may not provide the most efficient basis in any particular application, they do have the virtues of being (1) known, (2) somewhat related to the magnitudes of either the audited amounts or the differences, and (3) capable of implementation at a reasonable cost. When stratifying the sampling units, the auditor must choose (1) the number of strata, (2) the location of strata boundaries, and (3) the method of allocating the sample to the strata.

Number of Strata

Computer programs in current use allow either a fixed or a variable number of strata. In those allowing a variable number, the auditor may designate the number or else the computer program will iteratively test a number of alternatives and use the one that provides the smallest sample size. In some limited empirical work, it was found that using up to about five strata can be expected to result in large savings in sample size. With more strata, the incremental saving persists but becomes appreciably smaller because a few differences of larger size than anticipated can adversely affect the sample evaluation.

Frequently, out of the top stratum, 100 percent of the sampling units are selected and examined, because their large recorded amounts are of particular auditing concern. All individually significant items should always be audited. In this chapter, all the formulas presented for stratified sampling exclude consideration of such a top stratum. This means that the number of units in the top stratum must be added to the sample size determined from the formulas to obtain the total number of units to be examined by the auditor. It also means that the results concerning the audited amount for the top stratum must be added to the estimated audited amount computed from the other strata. Computing the achieved standard error or precision is not affected by the top stratum results because there is no contribution to the sampling error when a stratum is examined on a 100 percent basis.

Stratum Boundaries

There are several techniques for locating stratum boundaries. One of the simplest involves forming strata so that each contains approximately an equal recorded amount. To do this, the auditor first determines the top stratum (those units that will be selected and examined on a 100 percent basis), and then divides the remaining total recorded amount by the number of strata desired to obtain a target amount. The auditor then designates the stratum boundaries so that each stratum has nearly this target amount. This can be easily accomplished using the frequency analysis program described in chapter 9.

A more sophisticated but not necessarily better technique is in common use within the profession; it involves using the frequency analysis program in the following manner:

1. Determine the top stratum and exclude those units from further consideration.
2. Determine the number of strata desired and divide that number into the total cumulation of the square root of the frequency. This number becomes the target.
3. Assign boundaries so that each stratum has nearly the target value of the cumulative square root of the frequency.

For example, suppose the following represents the frequency distribution of recorded amounts for all but the top stratum.

Recorded Amt.	Frequency	Square Root of Freq.	Cumulative Square Root
0– 4.99	3464	58.9	58.9
5– 9.99	2516	50.2	109.1
10–14.99	2157	46.4	155.5
15–19.99	1581	39.8	195.3
20–24.99	1142	33.8	229.1
25–29.99	746	27.3	256.4
30–34.99	512	22.6	279.0
35–39.99	376	19.4	298.4
40–44.99	265	16.3	314.7
45–49.99	207	14.4	329.1

If four strata are desired, the total cumulative square root of 329.1 is divided by four to give 82.275 as the target value. Consequently, the upper boundaries should be as near as possible to the cumulative square roots of 82.275, 164.55, and 246.825. The actual closest boundaries to these are at the recorded amounts of 5, 15, and 25, so that the following represent the four strata.

Stratum	Frequency	Sum of Square Roots
0– 4.99	3464	58.9
5–14.99	4673	96.6
15–24.99	2723	73.6
24.99–49.99	2106	100.0

This is the best possible division that can be done with this frequency distribution using this technique, because the sum of square roots within each stratum is as close as possible to the target amount of 82.275. Since the frequency analysis program provides up to 200 cells, the achieved stratification can be closer to the target.[1]

Allocating the Sample to the Strata

A simple way of allocating the sample to the strata is to make the stratum sample size proportional to the total recorded amount within the stratum. Using this allocation scheme in conjunction with dividing the total recorded amount into equal portions produces nearly equal sample sizes for each stratum.

For example, suppose that 10,000 accounts have a total recorded amount of $4 million. Dividing these accounts into four strata so that each stratum contains about $1 million might produce the following results for a sample size of 300.

Stratum	Stratum Boundaries	Freq.	Total Rec. Amt.	Average Rec. Amt.	Stan. Dev. Rec. Amt.	Sample Size
1	0– 249	5500	$1,120,000	$ 203.64	$ 80	84
2	250– 749	3000	950,000	316.67	150	71
3	750–1499	1000	1,050,000	1050.00	200	79
4	1500–2999	500	880,000	1760.00	410	66
		10,000	$4,000,000			300

As can be easily observed, the larger recorded amounts in stratum 4 have a much greater representation in the sample than the smaller recorded amounts in stratum 1. The sampling proportion is .015 (84/5500) for stratum 1, .024 (71/3000) for stratum 2, .079 (79/1000) for stratum 3, and .132 (66/500) for stratum 4. Another way of describing this is to say that the sampling proportion is approximately proportional to the average recorded amount within each stratum, or that the stratum sample size is

1. Cells correspond to the division of the recorded amounts into intervals. When these cells have unequal lengths, a modification of this technique is necessary. The square root of the frequency is multiplied by the square root of the cell length before cumulating.

nearly proportional to the product of the number of stratum population items times the stratum average recorded amount.

Another allocation method widely used is known as *optimum* (or Neyman) *allocation*. The sample is allocated among the strata in proportion to the product of the number of stratum population items times the stratum standard deviation. This method also produces nearly equal stratum sample sizes when employed together with the cumulative square root technique of locating stratum boundaries. The term *optimum* reflects the fact that this allocation leads to the smallest possible standard error with respect to the standard deviation employed. When the stratification basis is the recorded amount, the resulting stratification is optimum only for the distribution of recorded amounts.

For example, if optimum allocation is used for the above data, the sample of 300 is allocated to the four strata as follows.

Stratum	Freq.	Std. Dev. Recorded Amts.	(Freq.) × (Std. Dev.)	Sample Size
1	5500	80	440,000	102
2	3000	150	450,000	104
3	1000	200	200,000	46
4	500	410	205,000	48
			1,295,000	

Using optimum allocation, the sampling proportion is .018 (102/5500) for stratum 1, .035 (104/3000) for stratum 2, .046 (46/1000) for stratum 3, and .096 (48/500) for stratum 4. The reason that the sample sizes vary so much among strata is that the boundaries were chosen to equalize the recorded amounts within each stratum rather than using the cumulative square root technique.

A third allocation method that can be used in some applications is to allocate the sample among the stratum in proportion to the number of stratum population items times the largest recorded amount within each stratum. For the above data, this results in the following allocation of the sample of 300.

Stratum	Freq.	Largest Rec. Amt.	(Freq.) × (Largest Amt.)	Sample Size
1	5500	247.65	1,362,075	62
2	3000	748.20	2,244,600	102
3	1000	1496.50	1,496,500	68
4	500	2950.10	1,475,050	68
			6,578,225	300

With this allocation method, the sampling proportion is .011 (62/5500) for stratum 1, .034 (102/3000) for stratum 2, .068 (68/1000) for stratum 3, and .136 (68/500) for stratum 4.

Relative Efficiency

Given that the auditor is restricted to using the recorded amounts as the basis for stratification, what method should be used to determine the number of strata, location of stratum boundaries, and allocation of the sample to the strata? Giving a definitive answer to this question does not seem possible based on the present state of knowledge. For those cases where the auditor does not anticipate a high proportion of differences in the population and where the differences are mostly overstatement errors, assigning approximately equal recorded amounts to each stratum and allocating the sample size in proportion to the recorded amount within each stratum is a simple but satisfactory procedure.

Using the cumulative square root of the frequency to locate the stratum boundaries together with optimum allocation produces a smaller sample size than dividing the population into equal dollar strata and allocating proportional to the stratum total recorded amount. However, the apparent superiority of the more complex method may be an illusion because the audited amounts may differ from the recorded amounts. For some patterns of differences between recorded and audited amounts, the simpler scheme may actually be better.

The broader question concerns the relative efficiency of recorded amounts as a stratification basis. In general, the effectiveness of the stratification depends upon the correlation between the recorded amounts and either the audited amounts or the difference amounts. When this correlation is uniformly high in each stratum, stratification based upon the recorded amounts produces sample sizes that are significantly smaller than unrestricted random sampling under all methods—mean, difference, ratio, and regression. This gain in efficiency is often enough to render the sample sizes capable of being used in practice.

Existence of numerous relatively large differences between audited and recorded amounts lowers the correlation. Consequently, the auditor must be careful in basing his sample size calculations solely on recorded amounts. The dangers are that a relatively small sample based on the presumed high correlation may be too small to reveal serious differences that exist or that large observed differences will adversely affect the evaluation.

To preclude this, the auditor might use the attribute tables as described in the previous chapter to determine the probability that a certain number of differences will be observed at various population occurrence rates. Using the attribute tables with a stratified sample is not strictly valid since the tables assume unrestricted random sampling. However, unless differences are more likely to occur in accounts with small balances than in accounts with large balances, the actual probability will be as large or larger than the tabled value.

To use the attribute tables, the auditor may equate the upper precision limit to the proportion of the population with differences between audited and recorded amounts. The reliability may then represent the probability of observing more than the tabled number of differences corresponding to a selected sample size and specified proportion of population differences. For example, the attribute tables based on the binomial show that observing nine occurrences in a sample of 100 gives an achieved upper precision limit of .14 with a reliability of .90. Reinterpreting, when the proportion of population differences is .14, there is a 90 percent probability of observing at least 10 differences in a sample of 100.

For a specified proportion of population differences, the auditor may use the tables to determine the sample size required to have a desired probability of observing a stipulated number of differences. For example, if the auditor believes that a population may have about 10 percent differences, a sample of 280 will afford a 95 percent chance of observing at least 20 differences. The required sample size for the same probability goes up to 400 when the proportion of population differences is only .07 and to 700 when the proportion of population differences is .04.

In the following sections are outlined the effects of stratification based on recorded amounts related to mean, difference, ratio, and regression estimation. Stratification affects the sampling distribution, the method of computing sample size, and the achieved precision of the resulting estimator.

Stratified Mean Estimation

The stratified mean estimate of the total audited amount is composed of the sum of the mean estimates over all strata. When there are L strata, this is represented as[2]

$$\hat{X}_{MS} = \sum N_i \bar{x}_i = N_1\bar{x}_1 + N_2\bar{x}_2 + N_3\bar{x}_3 + \cdot \cdot \cdot + N_L\bar{x}_L .$$

Within the ith stratum, N_i denotes the number of sampling units and $\bar{x}_i$ denotes the sample mean of audited amounts.

Sampling Distribution

Over all possible samples the mean of the stratified mean estimator equals the total audited amount. Thus, the stratified mean estimator is unbiased.

The estimated standard error of the stratified mean estimate equals the square root of the sum of the squares of the estimated standard errors within each stratum. In symbols,

$$\hat{\sigma}(\hat{X}_{MS}) = \sqrt{\sum N_i(N_i - n_i)\frac{S_{xi}{}^2}{n_i}},$$

where S_{xi} represents the estimated standard deviation of the audited amounts within the ith stratum. From the formula for the estimated standard error displayed in the mean estimation section of chapter 5, the estimated standard error for the ith stratum is

$$\frac{N_i S_{xi}\sqrt{1 - \dfrac{n_i}{N_i}}}{\sqrt{n_i}}.$$

Squaring this expression and rearranging terms gives the representation shown under the square root sign above.

The validity of precision and reliability statements depends upon the approximate normality of the sampling distribution. Neter and Loebbecke [18] found the sampling distribution to be approximately normally distributed for the populations in their study.

2. The subscript MS on $\hat{X}_{MS}$ denotes a *stratified mean* estimate of the total audited amount.

Determining Sample Size

The sample size required for a desired precision A and specified interval reliability R, is often determined by using the actual standard deviation of the recorded amounts in place of the estimated standard deviation of audited amounts. When optimum (or Neyman) allocation is used, the formula for this determination is as follows:

$$n = \frac{U_R^2(\sum N_i\sigma_{Yi})^2}{A^2 + U_R^2 \sum N_i\sigma_{Yi}^2}.$$

In this formula σ_{Yi} represents the standard deviation of recorded amounts in the ith stratum. Since all the recorded amounts are known, this can be computed rather than estimated. Also, the stratum standard deviation of recorded amounts is often a good approximation to the stratum standard deviation of audited amounts.

When the sample is allocated to the strata in proportion to the stratum total recorded amount, a different formula for determining the sample size is used. This formula is expressed as follows:

$$n = \frac{U_R^2 Y \sum N_i^2 \dfrac{\sigma_{Yi}^2}{Y_i}}{A^2 + U_R^2 \sum N_i\sigma_{Yi}^2}.$$

In this formula Y represents the total recorded amount in the population and Y_i represents the total recorded amount within the ith stratum.

To illustrate these formulas, suppose the auditor has a desired precision of \$150,000 at a .95 interval reliability ($U_R = 1.96$). Using the population data introduced earlier in this chapter, the required sample size for optimum allocation is as follows:

$$n = \frac{(1.96)^2[5500 \times 80 + 3000 \times 150 + 1000 \times 200 + 500 \times 410]^2}{(150{,}000)^2 + (1.96)^2[5500 \times 80^2 + 3000 \times 150^2 + 1000 \times 200^2 + 500 \times 410^2]}$$

$$= 275.66 \quad \text{or} \quad 276.$$

For the same requirements and data, the required sample size for allocation proportional to stratum total recorded amount is

$$n = \frac{(1.96)^2(4{,}000{,}000)^2\left[\dfrac{(5500 \times 80)^2}{1{,}120{,}000} + \dfrac{(3000 \times 150)^2}{950{,}000} + \dfrac{(1000 \times 200)^2}{1{,}050{,}000} + \dfrac{(500 \times 410)^2}{880{,}000}\right]}{(150{,}000)^2 + (1.96)^2[5500 \times 80^2 + 3000 \times 150^2 + 1000 \times 200^2 + 500 \times 410^2]}$$

$$= 310.25 \quad \text{or} \quad 311.$$

As expected, the optimum allocation method produces a smaller required sample size, but, as previously remarked, that fact alone does not make it superior in all circumstances.

Instead of using the standard deviation of recorded amounts, the auditor might take a preliminary sample to estimate the stratum standard deviations of audited amounts. From the standpoint of time and cost, the existence of the recorded amounts and the availability of a computer makes the use of a preliminary sample less desirable. Also, a preliminary sample is not always practical. In the case of a confirmation procedure, for example, the time required renders a preliminary sample impractical.

Evaluating Results

The achieved precision of the estimated total audited amount $\hat{X}_{MS}$, is estimated from the following formula:

$$A'_{MS} = U_R \sqrt{\Sigma\, N_i(N_i - n_i)\frac{S_{xi}{}^2}{n_i}}.$$

If the standard deviation of recorded amounts was used in determining the sample size, the achieved precision will equal the desired precision only when $S_{xi}{}^2 = \sigma_{Yi}{}^2$. This will rarely occur, since it requires both no observed differences as well as an estimated total audited amount equal to the total recorded amount.

Just as in the unstratified situation, the stratified mean estimate of the total audited amount is of little value to the auditor when no sample differences between audited and recorded amounts are observed. In that circumstance, the stratified mean estimate serves only to tell the auditor how representative his sample was. While this may be regarded as useful information, it says little about the reasonableness of the recorded amount.

The practical importance of stratified mean estimation is that in most situations the achieved precision (A'_{MS}) is as large or larger than the precision corresponding to any of the other techniques—difference, ratio, or regression. As a result, if the sample is planned so that A'_{MS} is a stipulated amount, the sample results can often be used to arrive at a useful audit conclusion. The possible use of the precision of the mean estimator in constructing a bound on overstatement error was discussed in the section on difference estimation in chapter 5.

Stratified Difference Estimation

The stratified difference estimate of the total difference amount is composed of the sum of the difference estimates over all strata. For L strata, the estimated total difference is as follows:

$$\hat{D}_S = \Sigma\, N_i\bar{d}_i = N_1\bar{d}_1 + N_2\bar{d}_2 + N_3\bar{d}_3 + \cdot\cdot\cdot + N_L\bar{d}_L .$$

The term N_i denotes the number of sampling units in the ith stratum and $\bar{d}_i$ denotes the mean difference in the ith stratum. The estimated total audited amount based upon the stratified difference estimator is given by the following:

$$\hat{X}_{DS} = Y + \hat{D}_S .$$

Sampling Distribution

Over all possible samples, the mean value of the stratified difference estimator equals the total population difference. Thus, the stratified difference estimator is unbiased.

The estimated standard error of the stratified difference estimate equals the square root of the sum of the squares of the estimated standard errors within each stratum. In symbols,

$$\hat{\sigma}(\hat{D}_S) = \sqrt{\Sigma\, N_i(N_i - n_i)\frac{S_{Di}{}^2}{n_i}},$$

where S_{Di} denotes the estimated standard deviation of the difference in the ith stratum. This formula also represents the estimated standard error of the estimated audited amount based on the stratified difference estimator $[\hat{\sigma}(\hat{X}_{DS})]$.

When the basis for stratification is the recorded amounts, Neter and Loebbecke [18] found the sampling distribution to be approximately normal when (1) the proportion of non-zero differences was at least 5 percent and the individual differences were relatively small in dollar amount or (2) when the proportion of differences was about 30 percent, and the individual differences were moderate or large in dollar amount.

The difficulty encountered for the proportion of differences between 5 and 30 percent for moderate and large differences may be related to the efficiency of the stratification based on recorded amounts as well as the relatively small sample sizes used in the study for populations where the difficulty was encountered.

Determining Sample Size

To determine the sample size required to achieve a desired precision at a specified reliability, the auditor would like an estimate of the standard deviation of differences within each stratum. While the auditor might take a preliminary sample within each stratum for that purpose, this is not ordinarily done because it is impractical. Instead, the standard deviation of recorded amounts within each stratum is used. If the number of strata is not too large, this procedure produces a larger sample than required whenever the correlation between audited and recorded amounts within each stratum exceeds one-half. Following this procedure, the sample size is determined by one of the formulas given in the section concerning stratified mean estimation.

When all or nearly all the differences can be expected to be overstatements, the sample size may be determined by adapting the approximation discussed in the unstratified case to the stratified situation. This involves conservatively estimating the proportion of non-zero differences within each stratum and following the suggestions made in the section on difference estimation in chapter 5 to calculate an approximation for the estimated standard deviation of difference amounts within each stratum. Denoting this approximation by $\hat{\sigma}_{Di}$, the sample size may be determined from the following formula when optimum allocation is used:

$$n = \frac{U_R^2(\sum N_i\hat{\sigma}_{Di})^2}{A^2 + U_R^2 \sum N_i\hat{\sigma}_{Di}^2}.$$

This, of course, is the same formula used for stratified mean estimation with $\hat{\sigma}_{Di}$ taking the place of σ_{Yi}, the standard deviation of recorded amounts. When the allocation is proportional to the stratum total recorded amount, the formula for the sample size is the same one used for stratified mean estimation with $\hat{\sigma}_{Di}$ replacing σ_{Yi}, that is

$$n = \frac{U_R^2 Y \sum N_i^2 \dfrac{\sigma_{Di}^2}{Y_i}}{A^2 + U_R^2 \sum N_i\sigma_{Di}^2}.$$

For example, using the data presented earlier in this chapter, the sample size has been previously computed as 276 for stratified mean estimation using optimum allocation. This same sample size may be used with stratified difference estimation. On the other hand, if the anticipated differences are mostly overstatements and the auditor conservatively believes about 10 percent of the recorded amounts could have differences, he can use the approximate method. If he expects each stratum to exhibit about the same proportion of differences, he uses $\hat{p}_D = .10$ for each stratum. This gives the following results for the approximate standard deviation of differences:

$$\hat{\sigma}_{D1} = \sqrt{.10 \times 80^2 + .10 \times .90 \times 203.64^2} = \$66.12.$$

$$\hat{\sigma}_{D2} = \sqrt{.10 \times 150^2 + .10 \times .90 \times 316.67^2} = \$106.18.$$

$$\hat{\sigma}_{D3} = \sqrt{.10 \times 200^2 + .10 \times .90 \times 1050^2} = \$321.29.$$

$$\hat{\sigma}_{D4} = \sqrt{.10 \times 410^2 + .10 \times .90 \times 1760^2} = \$543.69.$$

Using these values, the formula for the sample size corresponding to optimum allocation gives the following:

$$n = \frac{(1.96)^2[5500(66.12) + 3000(106.18) + 1000(321.29) + 500(543.69)]^2}{(150{,}000)^2 + (1.96)^2[5500(66.12)^2 + 3000(106.18)^2 + 1000(321.29)^2 + 500(543.69)^2]}$$

$$= 263.77 \text{ or } 264.$$

It is interesting to observe that while the sample sizes are not too different, the approximate standard deviations are lower for low-valued strata and higher for the high-valued strata. This reflects the fact that when a difference can be as large as the recorded amount, the variability among differences in high-valued strata is larger than the variability of the recorded amounts.

The allocation of the sample will be different under each of these methods. For example, if the sample size is increased to 300, the optimal allocation method applied to the approximate standard deviations ($\hat{\sigma}_{Di}$) gives the following results compared to the previously determined allocation based on the standard deviation of recorded amounts (σ_{Yi}).

Stratum	Allocation ($\hat{\sigma}_{Di}$)	Allocation (σ_{Yi})
1	85	102
2	75	104
3	76	46
4	64	46

Note that the allocation based on $\hat{\sigma}_{Di}$ is very close to that based on the stratum total recorded amount cited previously. This suggests that basing the total sample size on the standard deviation of recorded amounts (σ_{Yi}) and allocating the sample in proportion to the stratum total recorded amounts is a very good procedure when the stratum boundaries are set to include approximately equal recorded amounts within each stratum, and most of the differences are overstatements.

Finally, the auditor can use attribute tables as previously described to determine whether his computed sample size can be expected to produce a reasonable number of observed differences. For example, a sample of 264 has about a 90 percent probability of containing at least 20 differences when the proportion of accounts with differences is 10 percent (1,000 differences in this example). By increasing the sample size to 300, the probability of containing at least 20 differences is raised above 95 percent.[3]

Evaluating Results

The achieved precision of the stratified difference estimator ($\hat{D}_S$) (or the stratified difference estimate of the total audited amount ($\hat{X}_{DS}$)) at a specified reliability R is estimated by means of the following formula:

$$A'_{DS} = U_R \sqrt{\sum N_i (N_i - n_i) \frac{S_{Di}{}^2}{n_i}}.$$

where S_{Di} represents the estimated standard deviation of the population differences within the ith stratum based on the final sample results.

The number of differences required in order to obtain a good estimate of the achieved precision depends upon both the proportion and magnitudes of the differences. The most favorable situation is when (1) there are several observed differences within each stratum and (2) larger differences occur in the strata with larger recorded amounts.

When the total number of observed differences is small, the auditor can use a procedure similar to that described in the unstratified case to obtain an estimate of the standard deviation of stratum differences. This is described as follows:

1. From a binomial (or Poisson) table determine the achieved upper precision limit for the proportion of non-zero differences in the population. This upper precision limit ($P_U(m)$) is based on the total number of observed differences (m), the total sample size without regard to strata (n) and a desired one-sided reliability (R_1).

2. For the ith stratum, calculate the mean difference for only the sample units (m_i) containing non-zero differences, as follows:

$$\bar{d}_{m_i} = \frac{\sum d_{ij}}{m_i}.$$

3. For the ith stratum, calculate the estimated standard deviation for the non-zero stratum differences as follows:

$$S_D(m_i) = \sqrt{\frac{\sum d_{ij}{}^2 - m_i \bar{d}_{m_i}{}^2}{m_i - 1}}.$$

4. Determine the estimated standard deviation of the ith stratum as follows:

$$\hat{\sigma}_D(i) = \sqrt{P_U(m) S_D{}^2(m_i) + P_U(m)(1 - P_U(m)) \bar{d}_{m_i}{}^2}.$$

The resulting number for each stratum can be used in place of S_{Di} in computing the estimated achieved precision.

When only a few differences are observed within each stratum, the reliability factor used to estimate the achieved precision should not be taken from the normal table. Instead, the factor should be taken from the Student's t-table. In addition to the desired reliability, the auditor needs to specify the appropriate "degrees of freedom" when entering that table.

---

3. As stated previously, using the attribute tables with a stratified sample is not strictly valid because the tables assume unrestricted random sampling. However, unless differences are more likely to occur in accounts with small recorded amounts than in the larger recorded amounts, the actual probability will be at least as large as the tabled value.

Determining an appropriate number for the degrees of freedom is not a simple task. In his book, Cochran [5] presents a formula for approximating the degrees of freedom, but it assumes the underlying data are normally distributed and tends to overestimate the effective degrees of freedom under circumstances likely to occur in accounting populations. A reasonable number of the degrees of freedom would be the total number of observed differences minus the number of strata.

Using this procedure still requires observing a few differences within each stratum. In addition, the auditor should be reasonably satisfied that the observed differences are representative as discussed in chapter 5. When these conditions are satisfied the procedure described above accounts for the possibility that the occurrence rate of errors in the population could be larger than the fraction of errors observed in the sample.

When either no differences or only a few differences are observed, the auditor might consider determining a bound on the overstatement error as described in chapter 5. One such bound that requires no assumption concerning the approximate normality of the sampling distribution is available provided the sample has been allocated to the strata in proportion to the number of stratum population items times the largest recorded amount within each stratum (the third method described earlier). To compute this bound, the auditor determines the upper precision limit ($P_U(m)$) based on the total number of observed differences (m), the total sample size without regard to strata (n) and a desired one-sided reliability (R_1). The upper bound on overstatement error is then equal to the following:

$$P_U(m) \sum N_i \times (\text{Largest recorded amount in stratum } i).$$

For example, suppose the third allocation method described earlier is used to obtain a stratified sample of 300 for the data described and no differences are observed. The achieved upper precision limit is .01 at .95 one-sided reliability. The sum of the number of stratum population items times the largest recorded amount within the stratum has been computed as \$6,578,225, and so the bound equals

$$\$65{,}782 = .01 \times \$6{,}578{,}225.$$

Another bound that does depend on the approximate normality of the sampling distribution is to use the stratum standard deviation of recorded amounts (σ_{Yi}) in place of S_{Di} when using the formula for computing the achieved precision (A'_{DS}). This substitution will produce a conservative bound on the amount of overstatement as long as the square of the stratum coefficient of variation ($\sigma_{Yi}/\bar{Y}_i$) exceeds the proportion of non-zero differences within the stratum. In practice, the auditor can determine $P_U(m)$ as described above and use the bound as long as $(\sigma_{Yi}/\bar{Y}_i)^2 > P_U(m)$.

For example, for the same data, the squares of the coefficients of variation are as follows.

Stratum	Average Recorded Amount	Standard Dev. Recorded Amt.	Coefficient of Variation	Square of CV
1	\$ 203.64	\$ 80	.39	.15
2	316.67	150	.47	.22
3	1050.00	200	.19	.04
4	1760.00	410	.23	.05

If no differences are observed in the sample of 300, the achieved upper precision limit as cited above is .01 at .95 one-sided reliability. Therefore, the bound on the overstatement can be computed as follows, assuming that the sample was allocated to the stratum in proportion to the total recorded amount within the stratum (the first method described above) and using a reliability factor of 1.65 corresponding to a one-sided reliability of .95:

$$1.65\sqrt{5500 \times 5416 \times \frac{(80)^2}{84} + 3000 \times 2929 \times \frac{(150)^2}{71} + 1000 \times 921 \times \frac{(200)^2}{79} + 500 \times 434 \times \frac{(410)^2}{66}}$$

$$= 1.65 \times \$77{,}931$$

$$= \$128{,}586.$$

To be useful, the auditor should plan the sample so that when few or no differences are observed the magnitude of the bound is a desired size. When the latter method is to be used, the sample size can be determined using the formula for the stratified mean estimator augmented by the attribute tables. The attribute tables can be used to make sure that the sample size is large enough so that when a few errors are observed, the resulting upper precision limit on the proportion of errors in the population is smaller than the square of the stratum coefficients of variation for each stratum.

Stratified Ratio Estimation

The stratified ratio estimator most commonly used in auditing is known as the *combined* ratio estimator. The combined ratio estimator of the total audited amount is formed by the ratio of the stratified mean estimator of the audited amounts to the stratified mean estimator of the recorded amounts, multiplied by the total recorded amount. In symbols,

$$\hat{X}_{RC} = \left(\frac{N_1\bar{x}_1 + N_2\bar{x}_2 + N_3\bar{x}_3 + \cdots + N_L\bar{x}_L}{N_1\bar{y}_1 + N_2\bar{y}_2 + N_3\bar{y}_3 + \cdots + N_L\bar{y}_L}\right) Y$$

$$= \left(\frac{\sum N_i\bar{x}_i}{\sum N_i\bar{y}_i}\right) Y.$$

The combined ratio itself is denoted by the symbol $\hat{R}_C$. It may also be computed as one plus the ratio of the stratified difference estimator to the stratified mean estimator of the recorded amounts; that is,

$$\hat{R}_C = 1.0 + \frac{N_1\bar{d}_1 + N_2\bar{d}_2 + N_3\bar{d}_3 + \cdots + N_L\bar{d}_L}{N_1\bar{y}_1 + N_2\bar{y}_2 + N_3\bar{y}_3 + \cdots + N_L\bar{y}_L}$$

$$= 1.0 + \frac{\sum N_i\bar{d}_i}{\sum N_i\bar{y}_i}.$$

This also provides an estimate of the total difference between audited amounts and recorded amounts by letting

$$\hat{D}_{RC} = (\hat{R}_C - 1)Y.$$

Another type of stratified ratio estimate is based on what is called the *separate* ratio estimator. As the name implies, the separate ratio estimator is the sum of each stratum's ratio times its population recorded amount Y_i. The separate ratio estimator is more appropriate when the stratum ratios vary widely and this condition may occur in some auditing applications. However, the combined ratio estimator does not require observing several differences within each stratum as does the separate ratio estimator. For this reason, the separate ratio is not discussed further here, but additional details are presented in Appendix 7.

Sampling Distribution

Over all possible samples the mean value of the combined ratio estimate of the total audited amount does not equal the total audited amount, but the bias is negligible when the sample size is large enough to make the coefficient of variation of the stratified mean estimator of recorded amounts less than 10 percent. See Cochran [5], p. 169. Whenever planning is done on the basis of recorded amounts, and the planned precision expressed as a fraction of the recorded amount (the relative precision) does not exceed $.10U_R$, this condition will be satisfied. For example, when the interval reliability is .95 ($U_R = 1.96$), the coefficient of variation will be less than 10 percent whenever the planned relative precision is less than 20 percent.

The estimated standard error of the combined ratio estimate of the total audited amount is given by the following formula:

$$\hat{\sigma}(\hat{X}_{RC}) = \sqrt{\sum N_i(N_i - n_i)\frac{S_{RCi}^2}{n_i}},$$

where S_{RCi} denotes the estimated standard deviation of the population ratios within the ith stratum.

The formula used to calculate the estimated stratum standard deviation of the population ratios (S_{RCi}) depends upon whether sample audited amounts or sample difference amounts are used. For audited amounts,

$$S_{RCi} = \sqrt{\frac{\sum x_{ij}^2 + \hat{R}_C^2 \sum y_{ij}^2 - 2\hat{R}_C \sum x_{ij}y_{ij}}{n_i - 1}},$$

where the first subscript indicates stratum i and the second indicates the jth sample item within that stratum. For difference amounts,

$$S_{RCi} = \sqrt{\frac{\sum d_{ij}^2 + (\hat{R}_C - 1)^2 \sum y_{ij}^2 - 2(\hat{R}_C - 1)\sum d_{ij}y_{ij}}{n_i - 1}}.$$

Neter and Loebbecke [18] found that the sampling distribution of the combined ratio estimator was approximately normal under conditions similar to those pertaining to the stratified difference estimator.

One explanation for the results being so similar is that the ratio and difference estimators themselves are quite similar when the stratification is based on the recorded amounts. Comparing the formulas for $\hat{D}_{RC}$ and $\hat{D}_S$ shows that if $\sum N_i\bar{y}_i$, the denominator for $\hat{R}_C$, were exactly equal to Y, the formulas would become identical. When planning is done on the basis of the recorded amounts, the relationship between Y and $\sum N_i\bar{y}_i$ is controlled and any difference between them is less than A, the planned precision. Since relative precision is often taken as less than 10 percent, this means that the two estimators are also within the same 10 percent.

Determining Sample Sizes

The formula for determining the sample size required to achieve a desired precision at a specified reliability based on the optimum allocation using the combined ratio estimator is as follows:

$$n = \frac{U_R{}^2(\sum N_i S_{RCi})^2}{A^2 + U_R{}^2 \sum N_i S_{RCi}^2}.$$

Similarly, the sample size formula when the allocation is proportional to the stratum total recorded amount is as follows:

$$n = \frac{U_R{}^2 Y \sum \frac{N_i{}^2 S_{RCi}^2}{Y_i}}{A^2 + U_R{}^2 \sum N_i S_{RCi}^2}.$$

Again, these formulas are very similar to the corresponding formulas for both stratified mean estimation and stratified difference estimation. To use these formulas, the auditor needs to know the estimated standard deviation of population ratios within each stratum, S_{RCi}, and that information is not usually available unless a preliminary sample is feasible.

A frequently used procedure is to substitute the standard deviation of the stratum recorded amounts (σ_{Yi}) for the unknown (S_{RCi}). When the number of strata is not too large, this produces a larger sample than required if the correlation between audited and recorded amounts within each stratum exceeds one-half.

As was true with stratified difference estimation, when all or nearly all of the differences can be expected to be overstatements, the approximation discussed in the unstratified case can be adapted to this situation. This involves conservatively estimating the proportion of non-zero differences within each stratum and following the procedure described in the ratio estimation section of chapter 5 to obtain a bound for the estimated standard deviation of population ratios within each stratum. These numbers then take the place of S_{RCi} in the above formula.

To illustrate these procedures, consider once again the data presented at the beginning of the chapter. If the standard deviation of stratum recorded amounts, (σ_{Yi}) is used in place of S_{RCi}, the sample size will be the same as for stratified mean estimation. When optimum allocation is used, the sample size has been computed to be 276.

Adapting the approximation method discussed in chapter 5 to the stratified case, the auditor first determines a conservative proportion of anticipated differences within each stratum. As explained in the discussion of the stratified difference estimate, if the auditor believes about 10 percent of the recorded amounts could have differences and expects each stratum to show about the same proportion, he would set $\hat{p}_D = .10$. Representing the approximation for a stratum by $\hat{\sigma}_{RCi}$, the results are

$$\hat{\sigma}_{RC1} = \sqrt{.10 \times .90(80^2 + 203.64^2)} = \$65.64.$$

$$\hat{\sigma}_{RC2} = \sqrt{.10 \times .90(150^2 + 316.67^2)} = \$105.12.$$

$$\hat{\sigma}_{RC3} = \sqrt{.10 \times .90(200^2 + 1050^2)} = \$320.66.$$

$$\hat{\sigma}_{RC4} = \sqrt{.10 \times .90(410^2 + 1760^2)} = \$542.14.$$

Using these values, the formula for the sample size corresponding to optimum allocation gives the following:

$$n = \frac{(1.96)^2[5500 \times 65.64 + 3000 \times 105.12 + 1000 \times 320.66 + 500 \times 542.14]^2}{(150{,}000)^2 + (1.96)^2[5500 \times 65.64^2 + 3000 \times 105.12^2 + 1000 \times 320.66^2 + 500 \times 542.14^2]}$$

$$= 260.9 \text{ or } 261.$$

Comparing this result to the corresponding result for the stratified difference estimator shows very close agreement. For practical purposes, it is not necessary to do both, and, since the sample size for the difference estimator is always somewhat larger, that is the method used in the computer programs.

Besides using a formula to calculate the required sample size, the auditor can use the attribute table as previously described to determine the probabilities of observing a stipulated number of non-zero differences.

Evaluating Results

The achieved precision of the combined ratio estimator at a specified reliability R is estimated by means of the following formula:

$$A'_{RC} = U_R \sqrt{\sum N_i(N_i - n_i)\frac{S^2_{RCi}}{n_i}},$$

where S_{RCi} is the estimated standard deviation of the population ratios in stratum i.

One of the advantages of the combined ratio estimator is that the requirements concerning the number of observed differences is less stringent than for the stratified difference estimator. In particular, the estimate is not particularly sensitive to having some strata with no differences.

On the other hand, the behavior of the estimator closely parallels that of the stratified difference estimator, so the stability of the estimated precision depends upon both the frequency and magnitudes of the differences. In general, using the combined ratio in populations where the fraction of differences is smaller than about 5 percent is dangerous.

When several differences are observed, but the number is small, say less than 20, the estimated standard deviations (S_{RCi}) may underestimate the true stratum standard deviations. Adapting the procedure suggested in chapter 5 for the unstratified ratio estimation to the stratified situation yields the following:

1. From a binomial (or Poisson) table, determine an achieved upper precision limit ($P_U(m)$) based upon the total number of observed differences (m), the sample size (n), and a desired one-sided reliability (R_1).
2. For stratum i, if m_i is greater than one, let

   $$S^2_{RCi}(m_i) = \frac{n_i - 1}{m_i - 1} S^2_{RCi}.$$

3. Determine the estimated standard deviation of the stratum ratios as follows:

   $$\hat{\sigma}_{RCi} = \sqrt{P_U(m) S^2_{RCi}(m_i)}.$$

   Use $\hat{\sigma}_{RCi}$ in place of S^2_{RCi} in the formula for determining the precision.

Additionally, the achieved precision A'_{RC} can be computed using a reliability factor from the Student's t-table with degrees of freedom equal to the total number of observed differences minus the number of strata instead of using U_R based on the normal tables.

As before, this procedure only accounts for the possibility that the fraction of errors observed in the sample is smaller than the occurrence rate of errors in the population. This means that the auditor needs to be reasonably satisfied that the observed errors are representative of those that could occur in the particular situation. If he is not satisfied concerning their representativeness, he might consider enlarging the sample or using one of the procedures for obtaining an upper bound on overstatement error described in the section on stratified difference estimation. One of these procedures might also be used when no differences or only a few differences are observed.

Stratified Regression Estimation

Similar to stratified ratio estimation, the stratified regression estimator most often used in auditing is known as the combined regression estimator. The combined regression estimator uses the stratified mean estimators of the audited amount and the recorded amount in the following formula:

$$\hat{X}_{GC} = \sum N_i \bar{x}_i + b_c (Y - \sum N_i \bar{y}_i),$$

where b_c represents the estimated combined regression coefficient. The formula for the combined regression coefficient is complex. It is represented by the following:

$$b_c = \frac{\sum N_i (N_i - n_i) \dfrac{S_{XYi}}{n_i}}{\sum N_i (N_i - n_i) \dfrac{S_{Yi}{}^2}{n_i}},$$

where for the ith stratum,

$$S_{XYi} = \frac{\sum y_{ij}{}^2 - n_i \bar{y}_i{}^2}{n - 1}$$

and

$$S_{Yi}{}^2 = \frac{\sum y_{ij}{}^2 - n_i \bar{y}_i{}^2}{n_i - 1}$$

(the square of the estimated standard deviation of recorded amounts).

Just as for the unstratified case, this computation may be considerably simpler if the observed differences are used ($d_{ij} = x_{ij} - y_{ij}$):

$$b_c = 1 + \frac{\sum N_i (N_i - n_i) \dfrac{S_{DYi}}{n_i}}{\sum N_i (N_i - n_i) \dfrac{S_{Yi}{}^2}{n_i}},$$

where $S_{DYi} = \dfrac{\sum d_{ij} y_{ij} - n_i \bar{d}_i \bar{y}_i}{n_i - 1}$.

Sampling Distribution

Over all possible samples, the mean value of $\hat{X}_{GS}$ is not equal to the total audited amount but like the ratio estimator, the bias is small when the sample size is large. As an operating rule, the sample size should be large enough to make the coefficient of variation of the stratified mean estimate of recorded amounts less than 10 percent. Just as described for the stratified ratio estimator, this condition will be satisfied when the sample size is determined using the formula for the stratified mean estimator, and the desired relative precision (A/Y) does not exceed $.10\,U_R$, where U_R is the reliability factor used.

The estimated standard error of the combined regression estimate of the total audited amount is represented by

$$\hat{\sigma}(\hat{X}_{GC}) = \sqrt{\sum N_i(N_i - n_i)\frac{S_{GCi}^2}{n_i}},$$

where S_{GCi} denotes the estimated standard deviation of the regression population within stratum i. The estimated standard deviation S_{GCi} may be computed from the following formula:

$$S_{GCi} = \sqrt{S_{Xi}{}^2 - 2b_c S_{XYi} + b_c{}^2 S_{Yi}{}^2},$$

where as before, $S_{Xi}{}^2$ is the square of the estimated standard deviation of audited amounts, $S_{Yi}{}^2$ is the square of the estimated standard deviation of recorded amounts, and

$$S_{XYi} = \frac{\sum x_{ij}y_{ij} - n_i\bar{x}_i\bar{y}_i}{n_i - 1}.$$

An equivalent formula that uses the difference instead of the audited amounts is given by

$$S_{GCi} = \sqrt{S_{Di}{}^2 - 2(b_c - 1)S_{DYi} + (b_c - 1)^2 S_{Yi}{}^2},$$

where $S_{Di}{}^2$ represents the square of the estimated standard deviation of differences and

$$S_{DYi} = \frac{\sum d_{ij}y_{ij} - n_i\bar{d}_i\bar{y}_i}{n_i - 1}.$$

While there have been no empirical studies of the sampling distribution of this quantity, the results should be quite similar to those for the combined ratio estimator.

Determining Sample Size

The formula for determining the sample size required to achieve a desired precision at a specified reliability based on the optimum allocation using the combined regression estimator is

$$n = \frac{U_R{}^2(\sum N_i S_{GCi})^2}{A^2 + U_R{}^2 \sum N_i S_{GCi}^2},$$

where S_{GCi} denotes the estimated standard deviation of the regression population within stratum i.

When the allocation is proportioned to the stratum total amount, the sample size formula is

$$n = \frac{U_R{}^2 Y \sum \frac{N_i{}^2 S_{GCi}^2}{Y_i}}{A^2 + U_R{}^2 \sum N_i S_{GCi}^2}.$$

When values for S_{GCi} cannot be obtained from a preliminary sample, the auditor may use the same alternative procedures as described for the combined ratio estimator. The most conservative procedure is to use the standard deviation of the recorded amounts (σ_{Yi}) in the ith stratum in place of S_{GCi}. This will always produce a larger sample size than necessary unless the correlation between recorded and audited amounts is zero.

The bounds obtained for the case when all or nearly all of the differences can be expected to be overstatements are nearly the same as those obtained for the combined ratio case. The numerical illustration cited for the combined ratio estimator is applicable here as well.

In addition to using a formula to calculate the required sample size, the auditor can use the attribute tables as previously described to determine the probabilities of observing a stipulated number of non-zero differences.

Evaluating Results

The achieved precision of the combined regression estimator at a specified reliability (R) is estimated by means of the following formula:

$$A'_{GC} = U_R \sqrt{\sum N_i (N_i - n_i) \frac{S_{GCi}^2}{n_i}},$$

where S_{GCi} is the estimated standard deviation of the regression population described earlier.

Similar to the combined ratio estimator, the combined regression estimator does not require observed differences within each stratum in order to obtain a good estimate of the achieved precision. However, when not many differences are observed (say, less than 20), the estimated standard deviation (S_{GCi}) may be too low. In this circumstance, the following procedure may help:

1. From a binomial (or Poisson) table determine the achieved upper precision limit ($P_U(m)$) based upon the total number of observed differences (m), the sample size (n), and a desired one-sided reliability (R_1).

2. Calculate the mean difference for each stratum containing differences by using the formula

$$\bar{d}_{m_i} = \frac{\sum d_{ij}}{m_i},$$

where m_i is the number of non-zero differences observed within the stratum.

3. For each stratum containing non-zero differences, calculate the estimated standard deviation for the non-zero stratum differences as follows:

$$S_{Di}(m_i) = \sqrt{\frac{\sum d_{ij}{}^2 - m_i \bar{d}_{mi}}{m_i - 1}}.$$

4. Determine the estimated standard deviation for any stratum containing non-zero differences as follows:

$$\hat{\sigma}^2_{GCi} = P_U(m)S_{Di}{}^2(m_i) + P_U(m)(1 - P_U(m))\bar{d}_{m_i}{}^2 - 2(b_c - 1)S_{DYi} + (b_c - 1)^2 S_{Yi}{}^2.$$

Use the calculated $\hat{\sigma}^2_{GCi}$ in place of the S^2_{GCi} to compute the achieved precision.

As an additional safeguard, the reliability factor can be selected from the Student's t-table with degrees of freedom equal to the total number of observed differences minus the number of strata instead of using U_R based upon the normal table.

The remarks relative to the appropriate use of such a procedure made for the combined ratio estimator pertain to the combined regression estimator as well.

Summary

Using any of the statistical techniques discussed in this chapter involves three phases: planning, execution, and evaluation. The aspect of planning discussed here pertains to specifying the number of strata, the location of the stratum boundaries, the allocation of the sample to the strata, and the determination of the sample size. The number of strata should probably be between five and ten after designating a top stratum to be sampled 100 percent. One possible criterion to use is that the square of the ratio of the standard deviation of recorded amounts to the mean recorded amount within any stratum should exceed the anticipated proportion of differences within the stratum whenever this proportion is small, say, less than .05.

Boundary location and allocation of the sample to the strata could follow either the optimal allocation method, using the square root of the cumulative frequency, or the simpler allocation, based on the recorded amounts within the strata when the strata are created to have approximately equal recorded amounts.

Determining sample size is a two-stage process. First, the auditor specifies a tolerable sampling risk, a tolerable risk of overauditing, and a material amount. These determine the desired precision and reliability. The formula for the determination depends upon whether the auditor elects to use the positive or the negative approach.

Once the desired precision and reliability have been determined, one of the several formulas is used to obtain a required sample size. Which formula the auditor uses depends upon his knowledge at the time. All the formulas have the same general form—the only difference among them is that each uses a different standard deviation. Using the standard deviation corresponding to the recorded amounts will usually result in a sample size that is large enough for any evaluation technique. If this is done, the auditor should consider using the simpler allocation scheme.

Another possibility is to specify the proportion of anticipated differences, and use the technique described for the difference method to select the sample size. Of course, when the auditor has information from prior work concerning the magnitudes of the stratum standard deviations, these can be used in the formulas.

In either case, the auditor should consult the attribute tables to determine whether the computed sample size is large enough to observe a reasonable number of differences.

The appropriate sample evaluation depends upon both the number of differences observed in the sample and whether the auditor is reasonably satisfied that they are representative of the errors likely to occur in the circumstances. As stated in chapter 5, there is no safe minimum number of differences. Nevertheless, using the formulas to calculate the achieved precision seems reasonable whenever the number of observed differences is as large as 20. When the number is somewhat smaller than this, the auditor might consider the approximate procedure, provided he is satisfied concerning the representativeness of the observed differences.

Whenever either no differences or only a few differences are observed, the auditor might consider enlarging the sample or determining a bound for the overstatement error. Two such bounds were described in the section on stratified difference estimation. If the auditor anticipates using one of these bounds, the sample should be planned so that the calculated bound is useful.

Computer programs are described in chapter 9 to help the auditor perform the calculations involved in planning and evaluating stratified sampling plans.

Sampling Proportional To Size

Selecting sample items with probabilities proportional to the recorded amounts (pps sampling) is an alternative to stratifying the population by recorded amounts. Both techniques give greater weight to items with large recorded amounts than to items with small recorded amounts. Selecting items with probability proportional to recorded amounts is somewhat simpler than stratified sampling. This, together with its use in populations where differences between audited and recorded amounts are rare, has made the selection technique increasingly popular in auditing practice.

In chapter 4, three methods for selecting a sample with probability proportional to the recorded amount were described. To use any of those selection methods, the population should be divided into positive balances, negative balances, and zero balances. Each of these is treated separately. The topics discussed here include choosing the sample size and evaluating the sample results. This is done both for populations with high error rates as well as those with low error rates. It is assumed throughout that all recorded amounts are positive and that the audited amounts are likewise positive or zero.

High Error Rate Populations

When the population error rate is expected to be high, the appropriate pps estimate of the total audited amount is given by the following formula:

$$\hat{X}_{\text{pps}} = Y \frac{1}{n} \sum \frac{x_j}{y_j}.$$

Calculating this estimate involves multiplying the total recorded amount (Y) times the average of the ratios between the sample audited amounts and the sample recorded amounts $(1/n) \sum (x_j/y_j)$.

Alternatively, the estimate can be calculated using the observed sample differences ($d_j = x_j - y_j$) by means of the following formula:

$$\hat{X}_{\text{pps}} = Y\left(1 + \frac{1}{n} \sum \frac{d_j}{y_j}\right).$$

The total difference can be estimated as follows:

$$\hat{D}_{\text{pps}} = \hat{X}_{\text{pps}} - Y = Y \frac{1}{n} \sum \frac{d_j}{y_j}.$$

Sampling Distribution. Over all possible samples the estimator $\hat{X}_{pps}$ has a mean value equal to the total audited amount; thus, the estimator is unbiased. The standard error of the estimated audited amount is estimated by

$$\hat{\sigma}(\hat{X}_{pps}) = \frac{YS_p}{\sqrt{n}},$$

where S_p represents the estimated standard deviation of the ratios.

The estimated standard deviation S_p can be calculated from the following:

$$S_p = \sqrt{\frac{\sum\left(\frac{x_j}{y_j}\right)^2 - \frac{1}{n}\left(\sum\frac{x_j}{y_j}\right)^2}{n-1}}.$$

Using the observed differences, the formula is written

$$S_p = \sqrt{\frac{\sum\left(\frac{d_j}{y_j}\right)^2 - \frac{1}{n}\left(\sum\frac{d_j}{y_j}\right)^2}{n-1}}.$$

Neter and Loebbecke [18] found that the sampling distribution was approximately normal when the number of differences in the population is not too small. The fraction of differences required depends somewhat on the magnitudes of the distribution of the differences, and consequently general rules are difficult to give. Nevertheless, using a sample size sufficiently large to produce about 20 observed differences represents a reasonable operating rule.[4]

Determining Sample Size. The formula for determining the sample size required to achieve a desired precision at a specified reliability is as follows:

$$n = \frac{Y^2 U_R{}^2 S_p{}^2}{A^2}.$$

Using this formula requires an estimate of the standard deviation of ratios (S_p).

A convenient approximation is available whenever all or nearly all of the differences between audited and recorded amounts are overstatements. In this case, the auditor can obtain a useful bound on S_p by assuming that the amount of overstatement equals the recorded amount whenever the recorded amount is not correctly stated. To calculate the bound, the auditor should conservatively estimate the proportion of non-zero differences in the sample, multiply this proportion by its complement, and take the square root of the resulting product. The resulting number represents a conservative approximation for the standard deviation of ratios (S_p) and may be used in the formula to obtain an appropriate sample size.

To illustrate using this technique, suppose the auditor wants to estimate the difference between the audited and recorded amount for an inventory caused by differences in quantities, prices, or extensions. A priced perpetual record of the inventory amounts is available, and the auditor anticipates that about 20 percent of the items

4. This suggestion is based on an analysis of the distribution of S_p and using the rule should produce an estimate S_p whose coefficient of variation is less than 20 percent.

may have some difference. The total recorded amount of the inventory is \$5 million and the desired reliability is \$200,000 at an interval reliability of .95. Since it is not feasible to take a preliminary sample to obtain a value for S_p, the auditor uses .16 (.2 × .8) as an approximation for S_p^2. This is appropriate because it is anticipated that most monetary errors will be overstatement errors.

Using the formula the auditor finds that

$$n = \frac{(5{,}000{,}000)^2 \times (1.96)^2 \times .16}{(200{,}000)^2}$$

$$= 384.16 \text{ rounded to } 385.$$

Because not all the observed errors will be as large as the recorded amount, the achieved precision should be smaller than \$200,000. The following year the auditor can use the sample results to obtain a more accurate value for S_p.

Evaluating Results. The estimate of the total audited amount equals the total recorded amount multiplied by the average of the sample ratios. In symbols,

$$\hat{X}_{\text{pps}} = Y \frac{1}{n} \sum \frac{x_j}{y_j}.$$

The achieved precision of this estimate is calculated by the formula

$$A_{\text{pps}} = \frac{U_R Y S_p}{\sqrt{n}},$$

where S_p is the estimated standard deviation of ratios.

Continuing the example, suppose the 385 sample observations contained 50 observed differences ranging in magnitude from a low of 5 percent of the recorded amount to a high of 100 percent. While in an actual application, there might be as many as 50 different ratios, it is supposed that the observed ratios in this example were as follows.

Differences as Proportion of Recorded Amounts	Number of Observations
−.05	14
−.10	8
−.15	4
−.20	3
−.25	1
−.40	2
−.50	8
−.70	2
−.90	1
−1.00	7
	50

The negative signs indicate that all the observed differences were overstatements. For this data, $\sum (d_j/y_j) = -17.05$, and $-17.05/384 = -.0444$. Consequently, the estimated total audited amount using the difference formula is as follows:

$$\hat{X}_{pps} = \$5{,}000{,}000\,(1.0 - .0444)$$
$$= \$4{,}778{,}000.$$

The precision of this estimate at an interval reliability of .95 is $83,409. This precision is determined by first calculating the estimated standard deviation (S_p):

$$S_p = \sqrt{\frac{11.4975 - .7590}{384}} = .167.$$

Using this,

$$A'_{pps} = \frac{1.96 \times 5{,}000{,}000 \times .167}{\sqrt{385}}$$
$$= \$83{,}408.9 \text{ rounded to } \$83{,}409.$$

The achieved precision was less than half the desired precision, and, consequently, the sample was larger than necessary. Had the auditor known that S_p was about .17, the computed sample size would have been

$$n = \frac{5{,}000{,}000^2 \times 1.96^2 \times (.17)^2}{200{,}000^2}$$
$$= 69.4 \text{ rounded to } 70.$$

However, this sample size (70) would likely contain only about nine differences based on the fraction of differences observed in the sample (.13 = 50/385). To obtain 20 differences the sample size should be about 155.

Low Error Rate Populations

Apparently the first published article describing the use of pps sampling for accounting populations with low error rates was written in Dutch by A. van Heerden [12]. The first widespread use in the United States of pps sampling in populations with low error rates was due to the efforts of K. Stringer who developed methods for determining sample size, selecting the sample, and evaluating the sample results. His technique for sample evaluation was quite novel, and several others have developed modified versions of his basic results.

In several papers, both published and unpublished, Anderson, Leslie, and Teitlebaum (see [3] and [20]) have coined the term DUS standing for *dollar unit sampling* to represent their development of Stringer's basic idea.

In a recent paper, Fienberg, Neter, and Leitch [11] have established a firm theoretical base for the types of evaluations used by Stringer and by Anderson, Leslie, and Teitlebaum. The theory developed by Fienberg, Neter, and Leitch demonstrates that for a low number of observed errors the evaluation technique used by Stringer gives a conservative result. Further research is necessary before it can be definitely asserted that Stringer's method always yields a conservative result—but at this point, that appears to be a very good bet.

Stringer's method for evaluating a pps sample involves determining an upper bound for the amount of overstatement at a specified reliability level. The upper bound depends upon the number and magnitudes of the monetary errors in the sample. It does not, however, result from calculating an estimated error together with the precision of that estimate. For this reason, the term *upper bound* is used rather than *upper precision limit.*

To describe the evaluation procedure, suppose the auditor has selected a pps sample of size n and observed zero monetary errors. Since each individual dollar of the total recorded amount (Y) had an equal chance of being selected, the sample may be regarded as an unrestricted random sample of n individual dollars from a population of Y dollars.[5] The attribute tables may then be used to determine an upper precision limit for the proportion of dollars containing errors when none is observed ($P_U(0)$). For a specified one-sided reliability (R_1), the total error in the population is less than or equal to $Y \times P_U(0)$.

For example, if a pps sample of 100 is selected from a population with a recorded amount of $2 million, and no errors are found, a .95 upper precision limit on the number of dollars in error is .03. The total error in the population does not exceed $60,000 (.03 × $2,000,000) with .95 reliability.

Suppose that one of the n sampled accounts has an error. Let y_1 denote the recorded amount of the account with an error and d_1 denote the magnitude of that error. One way of evaluating this occurrence would be to determine the achieved upper precision limit corresponding to one observed error in a sample of size n at a specified one-sided reliability R_1. Labeling this $P_U(1)$, a bound on the total overstatement error in the population would be $Y \times P_U(1)$. In the above example, the total error would not exceed $95,000 (.0475 × $2,000,000) with .95 reliability.

While valid, this bound is much too conservative because it completely ignores the magnitude of the observed error. Only when the observed error (d_1) equals the recorded amount (y_1), would this bound be appropriate.[6] The Stringer method applied to this case gives a bound equal to the following:

$$YP_U(0) + Y[P_U(1) - P_U(0)]\,\frac{d_1}{y_1}.$$

For example, if the one observed error was $5.00 ($d_1$) in an account recorded at $20.00 ($y_1$), the Stringer bound at .95 reliability would be as follows:

$$\$2{,}000{,}000 \times .03 + \$2{,}000{,}000 \times [.0475 - .03] \times \frac{5}{20}$$

$$= \$60{,}000 + \$8{,}750$$

$$= \$68{,}750.$$

Generalizing this to the case where k overstatement errors are observed involves the following. First, order the relative errors (d/y) so that d_1/y_1 is the largest relative error and d_k/y_k is the smallest. For a specified one-sided reliability (R_1) determine

5. Because the sampling unit may be regarded as an individual dollar, Anderson and Teitlebaum [3] coined the descriptive term *dollar unit sampling.*

6. Since differences have been defined as audited amount minus recorded amount, d_1 should equal $-y_1$. However it is common in this type of evaluation to reverse the signs and define d as the recorded minus audited amount.

the $(k + 1)$ upper precision limits $P_U(0), P_U(1), \ldots, P_U(k)$. The bound is then computed as follows:

$$Y\left[P_U(0) + (P_U(1) - P_U(0))\frac{d_1}{y_1} + (P_U(2) - P_U(1))\frac{d_2}{y_2} + \cdots + (P_U(k) - P_U(k-1))\frac{d_k}{y_k}\right].$$

Since the objective of the procedure is to obtain a bound on the overstatement error, any observed understatement error is treated separately. The Stringer method applied to the observed understatement errors calculates a lower bound for such errors in much the same way as the upper bound is calculated. Thus when g understatement errors are observed, the relative amounts are ordered from largest understatement error (d_1/y_1) to smallest understatement error (d_g/y_g). This means $d_1/y_1 \leq d_2/y_2 \leq \cdots \leq d_g/y_g$.[7] For a specified one-sided reliability, a lower bound for the total understatement error is as follows:

$$Y\left[P_L(1)\frac{d_1}{y_1} + (P_L(2) - P_L(1))\frac{d_2}{y_2} + \cdots + (P_L(g) - P_L(g-1))\frac{d_g}{y_g}\right],$$

where $P_L(0) = 0$ and $P_L(k)$ is the lower precision limit corresponding to k observed occurrences at the specified one-sided reliability (R_1).

To facilitate the computation of these bounds, Poisson tables have been used. In the Stringer method, the precision adjustment factors are taken from the Poisson table and equal the difference in the precision limits divided by the sample size. For example, for a sample of 100 at .95 reliability, $P_U(0) = .03$, $P_U(1) = .0475$ and so $(P_U(1) - P_U(0)) = .0175$. Multiplying these amounts by the sample size (100), the precision adjustment factor for zero errors is 3 $(.03 \times 100)$ and for one error is 1.75 $(.017 \times 100)$ at a reliability of .95.

Sampling Distribution. One of the virtues of the Stringer method is that it does not depend on the sampling distribution being closely approximated by the normal distribution. In fact, the Stringer method does not depend upon any assumptions regarding the distribution of monetary errors—it is distribution-free. This, of course, is a strong virtue for the method.

Determining Sample Size. Ordinarily the achieved bound on the monetary error is compared with an amount the auditor considers material. If the achieved bound is no greater than the material amount, the auditor decides that no material error exists. The risk of being incorrect in this decision is less than or equal to the complement of the one-sided reliability (R_1).

One common method of selecting an appropriate sample size is to determine the sample size required to make the upper bound equal to the material amount when no monetary errors are observed. For a specified one-sided reliability, this means determining n so that

$$P_U(0) = \frac{M}{Y},$$

7. Since the signs of the understatement differences are *negative*, the *smallest* understatement is the *largest* algebraically.

where M represents the material amount. For example, if $M = \$100{,}000$, $Y = \$5$ million, and $R_1 = .90$, the required sample size is 120, because the achieved upper precision limit is .02 (100,000/5,000,000) when 0 occurrences are observed.

The drawback of this method for determining the sample size is that when a monetary error is observed, the upper bound exceeds the material amount. As a result, the auditor must obtain additional information to be able to decide whether or not there is a material amount of error. When, in fact, there is not a material amount of error, the result is overauditing. Stringer and others have recognized this and suggested determining the sample size so that the upper bound with no observed errors is below the material amount. The question is, how much below?

Kaplan [15] addressed this question and suggested that the auditor should specify his tolerable risk of overauditing when the actual amount of error is some very small amount such as .002 (.2 percent of the recorded amount). He presented a table and outlined a procedure to be used to determine the sample size to control the sampling risk ($\beta = 1 - R_1$) for a prescribed material proportion (M/Y) and the risk of overauditing (α) also at a prescribed proportion (Q/Y), where Q represents a very insignificant amount of monetary error.

Using Kaplan's method for determining the appropriate sample size, the auditor anticipates the possibility that he might see some monetary error and that error might be as large as the recorded amount. Since the procedure is based on the possibility that the relative error (d/y) could be 100 percent, the sample sizes are often larger than would be required if the auditor knew that the maximum relative error were smaller. Unfortunately, a lower maximum relative error is seldom known. The tables and directions for their use is included in Appendix 6A.

Evaluating Results. Currently several methods are being used to evaluate the results of a pps sample. The common objective of each method is to calculate a useful upper bound for the amount of overstatement in the population at a specified one-sided reliability level. Current research by Fienberg, Neter, and Leitch [11] may lead to bounds that are smaller than those currently used in practice. Such improvements will increase the usefulness of the pps procedure to auditors.

The evaluation should be confined to the set of account balances or transactions from which the sample was selected. Those accounts or transactions that are either excluded from the sampled population or examined on a 100 percent basis should be separately treated. For example, in an accounts receivable application, zero and credit balances are treated separately from the positive balances. Also, when systematic sampling is used to select the pps sample, all accounts or transactions exceeding the total recorded amount (Y) divided by the sample size (n) are examined on a 100 percent basis.

When no differences are observed in the sample, the upper bound for overstatement equals the achieved upper precision limit ($P_U(0)$), at a specified one-sided reliability (R_1), multiplied by the total recorded amount of those accounts included in the sampled population. For example, suppose that an unrestricted random sample of 100 is selected with replacement from a population whose total recorded amount is \$2 million. No observed differences in a sample of 100 gives an achieved upper precision limit of .03 at .95 one-sided reliability. Therefore, the achieved upper bound for overstatement error is \$60,000 ($.03 \times \$2{,}000{,}000$).

A similar result would be achieved had the auditor used systematic selection with the same target sample size (100). In that case, using a skip interval of \$20,000, all recorded amounts greater than \$20,000 (\$2,000,000/100) would be examined on a 100 percent basis. If there were, say, five such accounts totalling \$150,000, the auditor would evaluate these five separately from those selected from the remaining \$1,850,000 (\$2,000,000 − \$150,000). The sample size from the \$1,850,000 would be either 92 or 93 because 1,850,000/20,000 = 92.5. Observing no errors in a sample of 92, the auditor would conclude that the achieved upper precision limit is .0321 at .95 one-sided reliability. Therefore, the achieved upper bound for overstatement error for the sampled population would be \$59,385 (.0321 × \$1,850,000). Had the sample been 93, the upper precision limit would be .0317 and the resulting bound \$58,645 (.0317 × \$1,850,000). To the upper bound, the auditor adds any monetary errors found among the five recorded amounts examined on a 100 percent basis.

When some sample differences are observed, the achieved upper bound equals the upper bound corresponding to no observed differences plus an amount that depends upon the number and magnitudes of the observed differences. If all observed differences are overstatement errors, the Stringer method computes the additional amount by ranking the relative errors from high to low, multiplying a precision adjustment factor times each of the relative errors, summing the products, and multiplying the sum by the recorded amount of the sampled population. Continuing the above example, suppose the observed differences are as follows.

Recorded Amount	Audited Amount	Difference	Relative Difference
\$5000	\$1000	\$4000	.80
400	350	50	.125
750	600	150	.20

The precision adjustment factors may be based either on the binomial distribution or closely approximated by the Poisson distribution. The Poisson factors may be found in the study by Giles Meikle cited in the Selected Bibliography.

The following tables illustrate the results both for the case where unrestricted sampling is used and where systematic sampling is used.

Adjustment Factor	Unrestricted Sample $n = 100$	Relative Difference	Product
1.74	.0174	.80	.0139
1.56	.0156	.20	.0031
1.45	.0145	.125	.0018
			.0188

Adjustment Factor	Systematic Sample $n = 92$	Relative Difference	Product
1.74	.0189	.80	.0151
1.56	.0170	.20	.0034
1.45	.0158	.125	.0020
			.0205

For the unrestricted sample, the addition to the upper bound is calculated as \$37,600 (\$2,000,000 × .0188) and the resulting upper bound for overstatement is \$97,600 (\$60,000 + \$37,600).

For the systematic sample, the addition to the upper bound is calculated as \$37,925 (.0205 × \$1,850,000) and the resulting upper bound for overstatement in the recorded amount of \$1,850,000 is \$97,310 (\$59,385 + \$37,925). To this would be added any errors found among the five recorded amounts greater than \$20,000.

If some of the observed differences are understatement errors, the Stringer method subtracts a lower bound for observed understatement errors from the upper bound of overstatement as previously calculated. The objective of this is to calculate an upper bound for the *net* overstatement error. For example, suppose that the following two understatement errors were observed in addition to the three overstatement errors in the previous example.

Recorded Amount	Audited Amount	Difference	Relative Difference
\$3500	\$4000	−\$500	−.143
200	300	− 100	−.500

Just as for the overstatement errors, the precision adjustment factors may be based either on the binomial distribution or closely approximated by the Poisson distribution.

The following tables illustrate the results both for the case where unrestricted sampling is used and where systematic sampling is used.

Adjustment Factor	Unrestricted Sample $n = 100$	Relative Difference	Product
.30	.0030	−.143	−.00043
.05	.0005	−.500	−.00025
			−.00068

Adjustment Factor	Systematic Sample $n = 92$	Relative Difference	Product
.30	.0033	−.143	−.00047
.05	.0005	−.500	−.00025
			−.00072

For the unrestricted sample, the lower bound for understatement is −\$1360 (\$2,000,000 × −.00068), and the resulting upper bound for net overstatement is \$96,240 (\$97,600 − 1360).

For the systematic sample, the lower bound for understatement is −\$1332 (\$1,850,000 × −.00072) and the resulting upper bound for net overstatement is \$95,978 (\$97,310 − \$1332).

The resulting upper bound for net overstatement should be viewed with some caution. Strictly speaking, the reliability attached to this bound is less than that used for the calculation of the separate bounds. How much less? The best answer to this question is that the exact reliability is somewhere between the reliability used to calculate the bounds (R_1) and ($2R_1 - 1$). Thus, in the above example, the exact reliability is no less than .90 ($2 \times .95 - 1 = .90$).

To be sure of having a reliability of .95 for the bound on net overstatement, the auditor should calculate the separate bounds using a reliability of .975 ($.95 = 2 \times .975 - 1$). Doing this may result in an upper bound that is larger than the upper bound for overstatement calculated at a reliability of .95. In that case, the better bound for net overstatement corresponds to neglecting any contribution from observed understatement errors.

For example, the unrestricted sample above produced a bound on net overstatement equal to $96,240. If this bound is considered as having .90 reliability, then the auditor could have neglected the understated errors observed in the sample and evaluated just the overstatement errors at .90 reliability. This would produce an upper bound of $80,600 in this case, and, since this is smaller than the bound of $96,240, the auditor would use $80,600 as the upper bound for overstatement.

Summary

When should the auditor use pps sampling? Some auditors seem to feel the answer is "always," while a more moderate viewpoint is that the auditor should consider using the most appropriate technique available. When the proportion of population units with monetary differences is expected to be small and the audit objective is to test for the possibility of a material overstatement, pps sampling is the best statistical technique. When the proportion of population units with monetary differences is not small, pps sampling may still be a good technique, but not necessarily the most appropriate.

7

Using Statistical Sampling in Auditing

Chapter 1 discusses the general auditing process, and chapters 2 through 6 describe relevant statistical principles and some useful techniques. This chapter combines auditing and statistics. The discussion emphasizes ways of using statistical techniques in the design of audit programs, the application of audit procedures, and the evaluation of the evidential matter of the sample.

The major objective of this chapter is to describe one possible way to integrate statistical sampling into the audit planning process. One of the purposes of planning is to reduce the audit risk to a tolerable level. Using statistical sampling contributes to this by allowing the auditor to control one type of risk—the sampling risk that occurs when only some items in an audit population are observed instead of all. Integrating statistical sampling into the audit planning process aids the auditor in his effort to control the sampling risk associated with forming an opinion on the financial statements.

One way to achieve an integrated plan is to establish a material amount of monetary error for each account balance or class of transactions and consider the risk that all substantive tests taken together would fail to detect a material monetary error. Likewise, the necessary compliance tests would be designed to limit the risk of unwarranted reliance on the set of pertinent accounting control procedures.

At the next level of the planning process, the auditor plans the details of the tests for substance and compliance. For those that are statistical, this entails selecting an appropriate statistical objective, determining an appropriate sampling unit and frame, determining the sample size, and deciding the selection method. Integrated planning at this level can be accomplished by considering alternative ways of achieving the same tolerable audit risk and selecting the alternative that is most economically feasible.

The result of this planning process is a tentative audit program. The program is tentative for two reasons. First, it is based on the auditor's review and preliminary evaluation of the system of internal accounting control procedures. During the course of any tests of compliance—both statistical and nonstatistical—the auditor may find conditions that necessitate altering the planned program.

The second reason for the tentative nature of this program is that information from some of the planned substantive tests may cause the auditor to revise the audit program. Consequently, the audit program actually followed may well differ from the initially planned program.

The following flowchart shows the major steps in the audit process.

The discussion in this chapter concerns the ways these steps are affected when at least some of the tests of details are based on a statistical sample. As indicated in the flowchart, an evaluation of all the substantive tests of detail related to a particular account balance or class of transactions is required. The basic question to be answered at that point is whether all the tests considered together support the position that the recorded amount is not materially in error. If they do, no further testing of details is required in this phase of the audit. If they do not, additional evidence is required, and the audit program is revised accordingly.

To consider all tests together when some tests are statistical and others are nonstatistical, the auditor can allocate the measure of materiality between the two types of tests. Furthermore, all statistical tests can be planned so that when considered together, they provide evidence concerning the possibility of a material monetary error at a tolerable level of sampling risk.

One of the objectives of planning is to control the total audit risk of missing a material amount of monetary error. Because the sampling and nonsampling aspects of audit risk may conservatively be regarded as *additive* for planning purposes, the auditor can decide the appropriate levels for the sampling risks with the knowledge that the total risk that a substantive test may fail to detect a material amount of error is approximately the sum of the following:

1. The nonsampling risk that the test would fail to detect a material amount of error even when applied to every detail of the particular account balance or class of transactions.

2. The sampling risk that the test would fail to detect a material error because it is restricted to a sample of details of the transactions or balances.

The subsequent discussion focuses on controlling the sampling risk as one component of audit risk.

Preliminary Design

The preliminary design of the tentative audit program for each class of transaction or account balance when statistical sampling is used can be described as follows:

1. Determine the audit objectives.
2. Determine a measure of materiality.
3. Identify the nature of the tentative set of substantive audit procedures to be used to obtain evidence.
4. Identify the pertinent accounting controls.
5. Determine the range of possible reliance on those controls.
6. Determine the extent and timing of substantive tests for various degrees of reliance.

FLOWCHART OF
THE AUDIT PROCESS

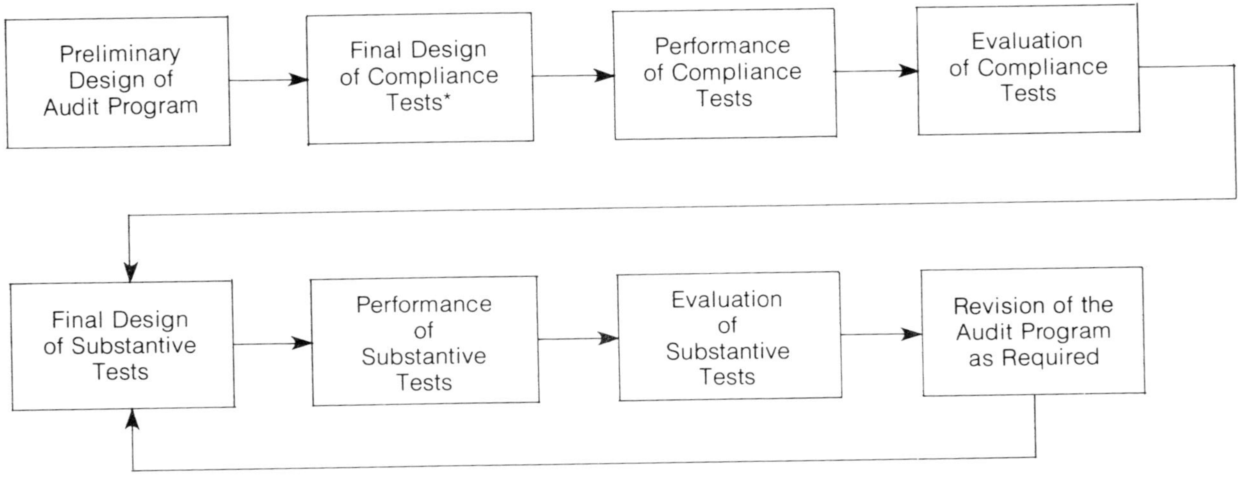

*If compliance tests are part of dual-purpose tests, this design includes the substantive procedures assuming that compliance is satisfactory. If internal accounting control is not relied upon, compliance tests are not required.

Exhibit 7.1

7. Determine the extent and timing of any required compliance test corresponding to each considered degree of reliance.
8. Select the degree of planned reliance.

Examining this list, the auditor notices two changes from what is customary when statistical samples are not used. First, the amount considered material for each account balance or class of transactions is to be explicitly stated, and second, the description suggests determining the extent and timing of the substantive and compliance tests under various degrees of reliance.

Stating the amount considered material is necessary for designing statistical tests of details that control the sampling risk of failing to detect a material error. On the other hand, if the auditor is accustomed to considering only the two extremes for the planned reliance (the minimum and maximum), there is no requirement that more be done when using statistical sampling. However, statistical sampling does afford the opportunity to determine the effect of intermediate degrees of reliance on the extent of the tests.

The following discussion of the preliminary design process focuses on how using statistical sampling affects the process.

Determine the Audit Objectives

Each statistical substantive test has both general and specific objectives. The general objective may be either deciding whether the amount of monetary error could be material (decision objective) or estimating the amount of monetary error (estimation objective).[1] The specific objective states in operational terms the types of monetary error to be examined in the test.

The choice of a general objective depends upon (1) the extent of error the auditor anticipates finding and (2) the costs and quality of alternative sources of additional information when that is required.

On the basis of preliminary evaluation of the internal accounting controls and any previous experience, the auditor may, for example, anticipate few if any monetary errors. In this case the decision objective would be appropriate. Either a positive or a negative approach may be used, as described in chapter 3. The description in this chapter will concentrate mainly on the negative approach.

When the auditor anticipates many monetary errors, either the decision objective or the estimation objective may be appropriate. In these circumstances, the estimation objective would be preferred if (1) the principal source of any additional information is additional sample items, and (2) it would not be feasible to obtain additional sample items at a later time. For example, the auditor may suspect that the accounts receivable have many errors and may want to be able to recommend an adjustment based on his sample. Since requesting confirmations on a second occasion would not be feasible, the auditor selects the estimation objective.

On the other hand, the auditor who suspects that many pricing errors are present in the inventory might select the decision objective, reasoning that if the sample indicates that there may be a material amount of error in the inventory, the sample can be expanded at a later date to obtain an estimate that could be used to support a sug-

---

1. Section 320B.25 describes the central importance of monetary error. Any procedure yielding an estimated audited amount also yields an estimated monetary error by simply subtracting the known recorded amount.

gested adjustment. Likewise, the auditor might select the decision objective if the inventory could be repriced in the event the sample indicated the possibility of a material error.

When the auditor is uncertain about whether monetary errors exist, the decision objective would seem to be more appropriate. As in the previous case, the only time the estimation objective would be better is when monetary error is expected and no opportunity exists for obtaining additional information, either from extending the sample or from other sources. When the auditor does not know what to expect and there is no other source of additional information, the decision objective should be used as the general objective, but the sample should be large enough to satisfy an estimation objective should an estimate be required.

After determining the appropriate general objective, the specific objective must be stated in operational terms. For example, in accounts receivable the specific objective might be to decide whether the recorded amounts contain material error with respect to existence, proper recording, and collectibility. A confirmation procedure may be used to obtain evidence pertaining to existence and proper recording, but not to obtain positive evidence concerning collectibility. The two procedures—requesting confirmation and testing collectibility—have the *joint* objective of deciding whether material error exists from the named causes. If both procedures are to be accomplished with the same sample, then the specific statistical objective may be deciding whether the difference between the recorded amount and the net realizable value is material. If collectibility is to be determined judgmentally or by using a separate statistical sample, the specific statistical objective of the confirmation procedure is limited to the gross receivable balance.

The importance of carefully stating the specific objectives cannot be overemphasized. A statistical sample of inventory items that is selected for the specific purpose of examining quantities, prices, and extensions does not yield any information concerning overstockage or obsolescence. Consequently, the specific objective cannot be stated as "decide whether the inventory amount is materially misstated" but must be limited to those sources of error that the auditor examines with the sample.

Materiality

The measure of materiality used to design the statistical tests of details pertaining to a particular account balance or class of transactions needs to be considered in relation both to other balances and classes of transactions and to other nonstatistical tests of details concerning the particular balance or class of transactions. Elliott [9] discusses this problem at some length and suggests a method for allocating the total measure of auditing materiality among the account balances and classes of transactions. Auditing materiality, according to Elliott, is related to the sensitivity of the audit to discovering monetary errors and, therefore, should not exceed the amount of monetary error the auditor deems material for the financial statements taken as a whole. The determination of materiality often contemplates such factors as net income, total assets, equity, and other considerations.

Elliott discusses allocating the overall material amount to the various account balances and classes of transactions. The scope of this chapter is more limited. The discussion here concerns allocating the material amount established for a particular account balance or class of transactions to the several tests the auditor may use. Thus, for the purposes of this chapter it is not necessary to follow any particular method for determining a material amount considered appropriate for a particular account balance or class of transactions.

Reliance

As discussed in chapter 1, the maximum degree of reliance on pertinent accounting controls ideally represents the auditor's assigned likelihood that the set of pertinent accounting controls would prevent or detect a material amount of monetary error. Minimum reliance corresponds to setting this likelihood equal to zero. The range of possible reliance would then correspond to numbers between 0 and the maximum degree. For example, if the auditor determined that the maximum degree of reliance was .7, corresponding to the judgment that assuming satisfactory compliance, the likelihood of occurrence of a material monetary error was .3, then the range of possible reliance would be from 0 to .7.

Since small variations in the degrees of reliance do not have much effect on the resulting extent of statistical tests, it is only necessary to select a few values from this range to express the possible degrees of reliance. For instance, the auditor might confine the possible degrees of reliance to 0, .3, .5, and .7 when the range is from 0 to .7. If the auditor does not want to use numbers, a qualitative scale may be substituted, such as none, little, moderate, and high, to express the possible degree of reliance.

The importance of analyzing each situation carefully and reaching a judgment based on the particular situation far outweighs the importance of being able to attach a precise number to the degree of reliance. Scoring schemes such as the one proposed by Elliott and Rogers [10] accomplish the objective without requiring the auditor explicitly to specify the likelihood that the set of pertinent accounting controls would prevent or detect a material amount of monetary error. However, decisions regarding the likelihoods are implicit in their proposal.

Extent of Substantive Tests

The following steps constitute a way of determining the extent of the statistical substantive tests for a particular account balance or class of transactions when both statistical and nonstatistical tests of details are used. Since timing decisions are no different whether or not statistical sampling is used, the determination of appropriate timing is not discussed.

1. Allocate the materiality amount between the statistical and any nonstatistical tests of details.
2. Determine the overall sampling risk.
3. Determine the overall risk of overauditing.
4. Determine the materiality, sampling risk, and risk of overauditing for each statistical sample.
5. Design each statistical sample.

Materiality. The materiality amount for the account balance or class of transactions is to be allocated between the statistical and nonstatistical tests. To accomplish this, the auditor judgmentally establishes an outside limit on the amount of monetary error that could remain undetected by the nonstatistical tests. This amount is then subtracted from the specified measure of materiality and the remaining amount is used in planning the statistical tests.

For example, for the tests of details of an inventory balance, suppose that quantities, prices, and extensions are to be tested statistically while obsolescence, overstockage, and cutoffs are to be tested nonstatistically. If $700,000 represents a material amount of error for all tests considered together, the auditor needs to decide how much of that amount to use in planning the statistical tests. This can be done by considering the outside limit for each of the nonstatistical tests. Thus, if, in the auditor's judgment, the planned tests of obsolescence and overstockage could each detect any misstatement exceeding \$150,000, while the test of cutoffs could detect a misstatement of \$100,000, the statistical tests of prices, quantities, and extensions could be planned using a material amount of \$300,000 (\$700,000 – \$150,000 – \$150,000 – \$100,000).

Sampling Risk. To determine a tolerable level of overall sampling risk, the auditor specifies a tolerable combined risk level $(1 - RS)$. This combined risk represents the sampling portion of the combined risk $(1 - R)$ described in chapter 1. The combined risk $(1 - RS)$ may be interpreted as the risk that the material amount of monetary error remains undetected because the auditor based conclusions on samples rather than auditing each individual balance contained in the account balance or every transaction. The level used by an auditor should be selected so that the sum of it plus the nonsampling portion of the combined risk represents a tolerable level for the overall audit risk $(1 - R)$.

For a given value of the sampling portion of the combined risk $(1 - RS)$, an appropriate value for the tolerable sampling risk can be determined by a formula that depends on the following:

1. C, the degree of reliance on internal accounting controls expressed in a scale between zero and less than one.
2. SP, the likelihood that analytical review procedures would detect a material misstatement if such existed.

This formula may be expressed as follows:

$$\beta = \frac{(1 - RS)}{(1 - C)(1 - SP)},$$

where β represents the sampling risk.

This formula is similar to the formula presented in section 320B.35. The chief difference between the two is that the factor $(1 - S)$ used in section 320B.35 represents the risk for all substantive tests—both tests of details and analytical review. In the formulation here, the risk for the substantive tests of details (β) has been considered separately from the risk for analytical review procedures $(1 - SP)$ so that the auditor may solve the equation for the tolerable sampling risk (β). The new factor (SP) is judgmentally determined by the auditor to represent the likelihood that the analytical review procedures would detect a material misstatement if such existed.

Algebraically, the two formulas are the same because the relationship

$$(1 - S) = \beta(1 - SP)$$

is appropriate whenever the two types of procedures may be regarded as independent. Thus, the formula presented here agrees with the one presented in section 320B.35 provided the nonsampling portion of the risk is negligible. In that case, of course, $(1 - RS)$ equals $(1 - R)$.

The following table illustrates how various combinations of judgmental reliance on internal accounting control (C) and of the judgmental likelihood (SP) that analytical review procedures would detect a material misstatement affect the sampling risk (β). Each combination of factors has the same sampling portion of the combined risk level ($1 - RS$) of approximately .05, which would be appropriate in practice provided the set of substantive procedures, if applied to every individual balance or transaction, would have a small risk of failing to detect the specified material amount of error (small nonsampling risk).

β	C	$(1 - C)$	SP	$(1 - SP)$	$1 - RS$
.05	0	1.00	0	1.00	.05
.10	0	1.00	.50	.50	.05
	.10	.90	.455	.555	.04995
	.20	.80	.375	.625	.05
	.30	.70	.296	.714	.04998
	.40	.60	.167	.833	.04998
	.50	.50	0	1.000	.05
.50	0	1.00	.900	.100	.05
	.05	.95	.895	.105	.049875
	.10	.90	.889	.111	.04995
	.20	.80	.875	.125	.05
	.50	.50	.80	.20	.05
	.60	.50	.75	.25	.05
	.80	.20	.50	.50	.05
	.90	.10	0	1.00	.05

The foregoing table reflects some of the different combinations of the various risks that lead to virtually the same combined risk level of .05. Notice that in the absence of any reliance on internal accounting controls ($C = 0$), and with no analytical review ($SP = 0$), the sampling risk (β) must be the same as the overall risk ($1 - RS$), which, in this table, is .05. Any combination of the complement of reliance ($1 - C$) and judgmental risk for analytical review procedures ($1 - SP$) whose product equals .50, allows a sampling risk (β) of .10. Likewise, any combination of the same factors whose product equals .10 permits setting the tolerable sampling risk (β) at .50.

The reliance on analytical review procedures (SP) is judgmentally determined. It represents the auditor's judgment concerning the likelihood that such procedures would detect a material monetary error in the account balance or class of transactions if such existed. Many auditors think that such procedures have only a moderate chance of disclosing material errors. In this book, the value of .33 ($SP = .33$) is used to illustrate such a moderate reliance.

Many auditors feel that if β is larger than .50, the value of a statistical substantive test of details is highly questionable. They reason that when a statistical sample has less than a 50–50 chance of indicating that there may be a material error when it exists, it is hardly worth doing. Consequently, when circumstances indicate a value of β larger than .50, the auditor could either omit the statistical sample and obtain the required satisfaction from other auditing procedures, or design the statistical sample to achieve a β equal to .50, and possibly reduce the scope of other auditing procedures.

The formula cannot be used directly unless the auditor specifies the internal control reliance on a scale between zero and less than one. However, it is still possible to specify an appropriate tolerable sampling risk that is consistent with the formula when both the degree of reliance and the effectiveness of analytical review procedures are expressed in qualitative terms. This is illustrated in the following table where the degree of reliance ranges from none to very high, and analytical procedures, if used, are judged to be only moderately effective. The sampling portion of the combined risk level ($1 - RS$) is maintained at .05.

Tolerable Sampling Risk

Reliance on Analytical Review	Reliance on Controls				
	None	Low	Moderate	High	Very High
None	.05	.07	.10	.15	.30
Moderate	.07	.10	.15	.25	.50

This table was constructed from the formula by assigning analytical review procedures (SP) either a 0 corresponding to no reliance or .33 corresponding to a moderate reliance. For reliance on controls (C), the correspondence was as follows: 0 representing no reliance, .3 representing low reliance, .5 representing moderate reliance, .7 representing high reliance, and .85 representing very high reliance. In some cases the sampling risk produced by the formula was modified to give a more convenient value.

Whether the formula or a table such as the one illustrated is used, the auditor can determine the tolerable sampling risk corresponding to selected degrees of reliance on internal accounting controls ranging from no reliance to the maximum possible reliance. Each of the values of the tolerable sampling risk can be used in the sample design to determine an appropriate sample size. Likewise, any required compliance tests can be designed for each selected sampling risk. Finally, the auditor can select the planned degree of reliance from among those considered as the one with the lowest cost in terms of the extent of the tests of details—both compliance and substantive.

Risk of Overauditing. While controlling sampling risk is necessary, the auditor may also control the risk that the statistical tests of details falsely indicate the presence of a potential material amount of monetary error. This risk is termed the *risk of overauditing* because when the statistical tests of details indicate the presence of a potential material error, the auditor needs to increase the audit scope.

Determining an appropriate value for this risk entails considering what would be done in the event the statistical tests of details indicate a potential material error. In some circumstances, increasing the extent of the tests of details might be required, while in others, additional procedures might be required. Not knowing the qualitative nature of any errors that might be discovered limits the auditor to planning alternative actions in general terms. Nevertheless, by contemplating the probable consequences, the auditor may be able to decide a reasonable value for the risk in a specific application.

Sample Design. The discussion of step 4 for determining the extent of substantive tests—determine the materiality, sampling risk, and risk of overauditing for each statistical sample—will come after the discussion of designing the statistical sample. Since step 4 is only required when there is more than one statistical substantive test of details for a particular account balance or class of transactions, the present discussion of designing the sample will cover the case of a single statistical test.

Sample design encompasses specifying an appropriate audit objective, identifying a sampling unit and frame, determining the sample size, and deciding on the selection method.

Audit objective—The objective of each statistical test of details should be clearly stated and coordinated with the audit objectives of the tests considered together. For example, the objective of a statistical price test of inventory may be to decide whether there is a stipulated material misstatement of the inventory balance caused by pricing errors. Note that the objective is limited to testing the effect of pricing errors only, and consequently the objective is not to determine whether the inventory balance might be materially misstated for other reasons.

Sampling unit and frame—The choice of a sampling unit and frame are taken together since the frame represents the listing of the sampling units. Many applications involve using a computer-based listing of the sampling units. In these cases, the frame and sampling unit are readily defined—the only requirement is to make sure the listing is complete. In some cases, the auditor can choose among alternative sampling units and the corresponding frames. For example, some accounting systems permit listing accounts receivable either by customer balance or by uncollected invoices. When alternatives exist, the auditor should consider the following characteristics:

1. Completeness—the frame corresponds closely to the population.
2. Efficiency—the frame and corresponding sampling unit afford both statistical and cost efficiency.
3. Convenience—the frame can be easily used.

Some examples of alternative frames and related sampling units are the following:

Inventory
- Listing of *inventory tags*
- Listing of *inventory products*
- Listing of *inventory locations*

Accounts receivable
- Listing of *customer balances*
- Listing of *uncollected invoices*

Determining sample size—Determining the appropriate sample size in a substantive test involves more than mechanically using a formula or a computer routine. Choosing one of the formulas presented in chapters 5 and 6 for determining a sample size to achieve the specified sampling risk and risk of overauditing relative to the determined material monetary amount, depends upon knowing that the corresponding evaluation technique can validly be used. The validity of many of the estimators depends upon the approximate normality of the sampling distribution and the stability of the estimated standard error. Whether these conditions are satisfied depends in turn on the proportion of sampling units with monetary errors and the magnitudes and algebraic signs of these monetary errors (that is, the difference, defined as the audited amount less the corresponding recorded amount). In many cases the auditor has very imprecise knowledge in this area. Consequently, the auditor strives to plan in such a way that regardless of the sample outcome, the data can be usefully analyzed. Although this cannot be guaranteed, careful planning will help achieve it.

Most of the time the auditor has the following information available when needed to determine the sample size: (1) the recorded amounts, and (2) the anticipated proportion of sampling units with monetary error. The sources of the auditor's information concerning the proportion of units with monetary error are (1) the preliminary evaluation of the internal accounting controls, and (2) experience from previous years or from a preliminary sample, if practicable.

Based upon this limited information the auditor needs to determine a sample size that will achieve the specified tolerable sampling risk (β) for a specified material amount (M) and the tolerable risk of overauditing (α).[2] Two basic methods of doing this will be discussed. Using one method, involving one of the standard estimators, the auditor translates the risk requirements (α, β) and the material amount (M) into a desired precision (A) stated either at an interval reliability (R) or a one-sided reliability (R_1). The interval reliability (R) is used to design the sample when following the positive approach and the one-sided reliability is used when following the negative approach. The equations used for this translation are as follows:

Positive Approach

$$A = \frac{z_{\alpha/2} M}{z_\beta + z_{\alpha/2}} \quad \text{and}$$

$$U_R = z_{\alpha/2},$$

$z_{\alpha/2}$ represents the normal factor corresponding to a risk of overauditing equal to α and z_β represents the normal factor corresponding to a sampling risk of β.[3]

Negative Approach

$$A = \frac{z_\beta M}{z_\beta + z_{\alpha/2}};$$

$$U_{R_1} = z_\beta.$$

As above, $z_{\alpha/2}$ represents the normal factor corresponding to a risk of overauditing equal to α and z_β represents the normal factor corresponding to a sampling risk of β.

The other basic method involves obtaining a useful upper bound to the monetary error and uses the specified risks (α, β) and the material amount (M) directly in a table

2. Using the symbols α for the risk of overauditing and β for the tolerable sampling risk corresponds to the use of these symbols in statistical hypothesis testing only when the positive approach is used. Nevertheless, they will also be used in describing the negative approach.
3. See chapter 3 for more details.

"look-up."[4] This method is appropriate for analyzing populations with a low incidence of monetary error.

Which method should the auditor select? It depends upon the anticipated incidence of monetary error. For a moderate or high incidence of monetary error, using one of the standard estimators is desirable and when the incidence is low, a bound is desirable. To translate this into terms the auditor can use, it is necessary to specify more exactly what constitutes a *low, moderate,* or *high* incidence of monetary error. For statistical sampling purposes, an anticipated occurrence rate is regarded as low if it is .05 or less, moderate if between .05 and .15, and high if .15 or more.

When the anticipated occurrence rate is low, the auditor expects to observe few if any monetary errors in the sample. Consequently, the auditor can determine the sample size to provide a useful upper bound on the monetary error. This can be done, as explained in chapter 6, using either a pps sample or a stratified random sample.

When the auditor anticipates that the incidence of monetary error will be higher than 5 percent, the choices for determining a sample size are broader. A pps sample has the advantage of providing an estimate of the total difference when there are many observed differences and providing an upper bound for the total difference when there are few observed differences. Determining an appropriate sample size involves considering both possibilities. Thus, the auditor should determine the sample size required by both the standard estimator and the procedure for low incidence of monetary error and use the larger of the two. Using the conservative approximation for the standard error of the pps estimator will ordinarily produce a sample size sufficiently large for the standard estimator.

The auditor may also anticipate using the ratio, difference, or regression estimator. One procedure for determining sample size that is used extensively is to stratify the frame by recorded amounts and use the formula appropriate for a stratified mean estimator based on the true standard deviation of the recorded amounts. When the number of strata is not too large, this will ordinarily produce a sample size larger than required, but the auditor should test the adequacy of the resulting sampling size by consulting the attribute tables as explained in chapter 6. Using these tables gives assurance of observing differences only under some assumptions concerning the incidence of population differences.

A modification of this procedure is appropriate when the auditor anticipates that nearly all the differences are overstatement errors. The procedures outlined in chapter 6 may be used to determine an appropriate sample size for this situation.

In summary, determining an appropriate sample size to meet the planned risk specifications involves anticipating what is likely to be observed. Moreover, the prudent auditor will plan so that when expectations are not met, the sample will still provide useful information. This may well require using a larger sample than would be necessary under the best of circumstances. Of course, when the cost of obtaining additional sample items is small, the auditor might choose to plan as if the most optimistic expectations would be met and then add more if required.

Selecting the sample—The foregoing discussion of determining sample size for a substantive test was limited to using either a stratified random sample or a pps sample. Selecting a sample when the frame is stratified by recorded amounts can be

4. Some sources refer to these upper bounds as upper precision limits, but that is not strictly correct. A precision limit is based on an estimate together with its achieved standard error. An upper bound need not be based on an achieved standard error. Consequently, not all upper bounds are upper precision limits.

The auditor should review each account balance or class of transactions to determine the most economically efficient approach in each case. Whatever is decided can be planned to control the sampling risk for either one or several statistical tests considered collectively.

Extent of Compliance Tests

The following steps represent a method of determining the extent of the statistical compliance tests corresponding to each selected degree of reliance. As in the case of substantive tests, decisions affecting the timing are no different when statistical sampling is used in place of a judgment sample.

1. Determine the audit objectives.
2. Determine the overall risk of unwarranted reliance pertaining to each planned substantive test.
3. Determine the overall risk of overauditing pertaining to each planned substantive test.
4. Determine the risks of unwarranted reliance and overauditing for each individual compliance test.
5. Design each statistical test of compliance.

The auditor plans each statistical test of compliance with a pertinent accounting control so that the risk of unwarranted reliance is controlled. Because unwarranted reliance increases the risk of failing to detect a material amount of monetary error, the auditor plans the statistical compliance tests in conjunction with the tentatively planned substantive tests. For each considered degree of reliance, the auditor determines the extent of the required compliance tests. A degree of reliance of 0, corresponding to no reliance, does not require any compliance tests. For degrees of reliance higher than 0, compliance tests are necessary.

Not all compliance tests can be based on a statistical sample. Segregation of duties does not always leave an audit trail of evidence and, consequently, as stated in section 320.59, "tests of compliance in these situations necessarily are limited to inquiries of different personnel and observation of office personnel and routines to corroborate the information obtained during the initial review of the system." This example illustrates the general situation: Those controls whose operation must be established through inquiry and observation are not subject to statistical sampling.

Other controls, such as approvals, "require inspection of the related documents to obtain evidence in the form of signatures, initials, audit stamps, and the like, to indicate whether and by whom they were performed and to permit an evaluation of the propriety of their performance.[10] This type of control can be tested using a statistical sample.

Audit Objectives. The audit objective of a compliance test is "to provide reasonable assurance that the accounting control procedures are being applied as prescribed."[11] This statement of the general audit purpose is further refined to state, "samples designed for this purpose should be evaluated in terms of deviations from,

---

10. Section 320.58.
11. Section 320.55.

or compliance with, pertinent procedures tested, either as to the number of such deviations or the monetary amount of the related transactions."[12]

The statistical objective of a compliance test, then, is to decide whether the deviations from pertinent procedures are too great to justify the planned reliance. Chapter 1 discusses an approach for determining for each pertinent control, or set of controls, the range of deviation that would be considered consistent with the planned reliance. Corresponding to this viewpoint, the statistical objective of a compliance test is to decide whether the rate of compliance deviations from a prescribed procedure or set of procedures could be as large as a determined threshold rate (P_0). As the previous quotation from section 320B.15 shows, this threshold rate may be expressed either as a proportion of the number of transactions or as a proportion of the monetary amount of the related transactions. However, even when expressed as a proportion of the number of transactions, it is necessary to obtain some indication of the magnitude of the *potential* monetary error.

Special care is required when the determined threshold rate applies to a set of related procedures as discussed in chapter 1. For example, if one pertinent procedure is designed to prevent an error while another is designed to detect and correct any error that occurs, the threshold rate (P_0) refers to the probability that a compliance deviation occurs for both procedures on the same transaction. When a sample is used to test the two together, this means defining an occurrence as failure to comply with both procedures.

When the set of related procedures jointly prevent an error from occurring, the threshold rate applies to the probability that there is a compliance deviation from any one of the set. Consequently, when a test is used, an occurrence is defined as a compliance deviation with any one of the set.

Overall Risk. The contribution to the overall audit risk arising from compliance tests is the risk of unwarranted reliance on internal accounting control. Unwarranted reliance may occur when the auditor incorrectly decides that the procedures are being followed to a satisfactory degree when, in fact, compliance deviations are more numerous than satisfactory. The result of this type of mistake is that the nature, extent, or timing of substantive tests is determined on misleading information with the consequence that the risk of missing a material amount of monetary error is greater than it ought to be.

The risk of unwarranted reliance has both statistical and nonstatistical aspects. The nonstatistical aspect pertains to the effectiveness of the audit procedures in detecting noncompliance, even if applied to every transaction. The statistical aspect pertains to that additional risk caused by the fact that the procedure was restricted to a sample of the transactions. As in the case of substantive tests, the two risks are approximately additive.

The statistical aspect of the risk of unwarranted reliance is measured by the probability that the auditor decides a particular rate of compliance deviation is satisfactory for the planned reliance when in fact the rate of compliance deviation is at the threshold rate of unsatisfactory compliance. Controlling the risk at the threshold rate means the risk is smaller should the actual rate exceed the threshold rate. When several control procedures are used to prevent or detect a particular type of error, the auditor needs to consider the set of procedures together in assigning an allowable sampling risk of unwarranted reliance.

---

12. Section 320B.15.

For each account balance or class of transactions, the overall sampling risk of unwarranted reliance may be determined to limit the auditor's sampling risk of missing a material amount of monetary error as described in chapter 1. The auditor might decide, for example, that should the compliance tests not corroborate the planned reliance, he would use a sampling risk of β_0 corresponding to no reliance. When the compliance tests incorrectly corroborate the planned reliance, the auditor incorrectly uses the planned (higher) sampling risk (β). The difference between these two levels of sampling risk ($\beta - \beta_0$) multiplied by the risk of unwarranted reliance (δ) represents the expected increase in his sampling risk when the rate of compliance deviation is at the threshold level and the monetary error equals a material amount (M). The auditor's judgment to limit this increase to, say, no more than .01, allows determination of the appropriate value for δ.

To illustrate, suppose the auditor uses a procedure for determining tolerable sampling risk as set forth in the Tolerable Sampling Risk table previously cited. The following table shows how the risk of unwarranted reliance may be determined if the additional sampling risk is allowed to be .01. As before, analytical review is assigned a value of .33 to represent the auditor's judgment that the review procedures have about a 1 in 3 chance of discovering a material monetary error.

	No Reliance on Analytical Review		
Degree of Reliance on Internal Control	Tolerable Sampling Risk	Increase	Risk of Unwarranted Reliance (added risk = .01)
None	.05		
Low	.07	.02	.50
Moderate	.10	.05	.20
High	.15	.10	.10
Very high	.30	.25	.04

	Moderate Reliance on Analytical Review		
Degree of Reliance on Internal Control	Tolerable Sampling Risk	Increase	Risk of Unwarranted Reliance (added risk = .01)
None	.07		
Low	.10	.03	.33
Moderate	.15	.08	.12
High	.25	.18	.05
Very high	.50	.43	.02

This table shows the effect of increasing the reliance on the pertinent accounting controls. At first glance, situations using a risk of unwarranted reliance equal to .50 or .33 apparently contradict the examples in section 320B.24 using risk levels no greater than .10. However, when the auditor considers that SAS no. 1 contemplates adjusting the threshold rate depending on the degree of planned reliance, while the procedure outlined here keeps that rate fixed, it will be seen that there is no contradiction.

For example, suppose the auditor determines that the threshold rate is .05 and the anticipated compliance deviations may be as high as .04. A sample of 100 would provide a risk of unwarranted reliance equal to about .44. The same sample size would provide a risk of unwarranted reliance of .05 at a threshold rate of .09. Consequently, keeping the risk at .05 and raising the threshold rate from .05 to .09 has the same effect as raising the risk from .05 to .45 and keeping the threshold rate the same. Justifying an increase in the risk of unwarranted reliance because the degree of reliance is less seems easier than justifying an increase of allowable compliance deviations when there is less reliance, but either method is clearly acceptable.

The outcome of this analysis is a determination of the allowable risk of unwarranted reliance corresponding to each considered degree of reliance and specified tolerable sampling risk. A similar type of analysis may be used when the planned substantive tests are nonstatistical even though numerical values may not be available. The key idea is that unwarranted reliance affects the risk of missing a material amount of error. Consequently, the auditor needs to consider both what would be done in the absence of reliance and what should be planned when the compliance tests corroborate the planned reliance. If the tentative audit program employs a relatively high risk of missing a material amount of monetary error, the compliance tests should provide a low risk of unwarranted reliance. If, on the other hand, the auditor considers the risk of missing a material amount to be moderate, a higher risk of unwarranted reliance may be used.

Individual Risk. Once the auditor determines an appropriate value for the risk of unwarranted reliance (δ) to be used for a particular account balance or class of transactions, the risk levels can be planned for testing compliance with each of the pertinent accounting controls. For each substantive test, only those controls that pertain to the types of errors that could be detected by the substantive test need be considered. For example, a statistical sample used to request confirmation of accounts receivable depends only on the controls pertaining to existence and accuracy of the recorded amounts. The pertinent controls may be those related to the following objectives:

1. Shipments are authorized and accurate.
2. Invoices are accurately prepared and are properly recorded.
3. Entries to accounts receivable (both control and detail) are authorized.
4. Cash receipts are adequately controlled and accurately recorded.

Suppose that each of the four sets of pertinent procedures is tested using the same risk of unwarranted reliance (δ) equal to, say, .05. The auditor decides that compliance is satisfactory only when it can be concluded that each set's rate of compliance deviation is below its threshold level. Conversely, the auditor decides compliance is not satisfactory whenever tests indicate that any of the rates of compliance deviation may be above the threshold level. Following this decision rule, the combined risk of unwarranted reliance is no greater than .05 whenever any set's rate of compliance deviation equals its threshold level. If more than one set's rate of compliance deviation equals its threshold rate, the risk of unwarranted reliance is much less than .05 (.0025 for two, .0001 for three, and nearly zero for all four).

There is, however, a disadvantage to using the same tolerable risk of unwarranted reliance for each compliance test. It may be possible that some combination of individual rates of compliance deviation could produce a potential material monetary error even though the rate for each set is below its threshold level. For example, some proportion of inaccurately prepared invoices together with some proportion of inaccurately recorded cash receipts could produce a material overstatement of accounts receivable. The risk of unwarranted reliance when the potential material error is caused by some combination may be well above the tolerable level.

Can this disadvantage be overcome? Probably, but the technology required to do it is quite complex, and further development is required before any practical solution can be offered. In the meantime, the auditor needs to be aware that present practice does not give much information regarding the possibility that material monetary errors may arise from a combination of causes.

After the auditor has determined the relevant sources of monetary error and the allowable risk of unwarranted reliance for each, consideration should be given to the pertinent procedure or set of procedures that are designed to prevent or detect each particular type of error. For example, the following procedures may be designed to achieve the objective that shipments are authorized and accurate:

1. Sales orders are maintained under numerical control.
2. Unfilled orders are periodically reviewed.
3. Sales orders are initialed as being approved by responsible employees.
4. Sales orders are checked for quantitative agreement with shipping orders and for accuracy of prices and extensions and are then initialed.

Compliance tests for a set of pertinent procedures should be planned so that the risk of unwarranted reliance when the set is evaluated as a single procedure is at the desired level.[13] When some of the procedures are designed to prevent the error from occurring (prevention controls) while others are designed to detect any errors that occur (detection controls), there are two ways of satisfying this requirement. The auditor may select a sample and define an occurrence as a lack of evidence of compliance with both types of procedures. In this case, the specified risk level would apply to the two procedures regarded as a single procedure. This is only possible when a statistical sample is used to test compliance with both the prevention controls and detection controls.

Frequently the prevention controls cannot be tested statistically because they rely on separation of duties or using prenumbered forms. In this case they must be tested separately from the detection controls. What risk of unwarranted reliance is appropriate for the statistical test of the detection controls? An answer is to use the planned tolerable risk of unwarranted reliance but to regard the combined threshold rate as a product of the individual threshold rates. For example, if the threshold rate for the controls considered together is .05 and the combined risk of unwarranted reliance is .10, the auditor may first judgmentally decide the likelihood that the prevention controls could allow a monetary error to occur. If this likelihood is taken as one, the auditor acts most conservatively by using the threshold rate of .05 and risk of unwarranted reliance of .10 for the compliance test of the detection controls. If the likelihood is only .25, the threshold rate for the compliance test of detection controls rises to .20 but the risk remains at .10.

---

13. This corresponds to section 320B.20.

When there are several pertinent procedures that constitute the set of controls, the requirement to evaluate the set as a single procedure means that an occurrence should be defined as evidence of noncompliance with any of the procedures. In the previous example, the first two pertinent procedures are to be tested nonstatistically (through inquiry and observation) while the latter two that leave documenting evidence are to be tested statistically. This means that the sample should be designed so that any deviations from the required evidence for approval or for accuracy checks can be evaluated at a specified risk level. This simply requires that a common sample be selected for both attributes. The auditor may want to examine each attribute separately, but that is not required for corroborating the planned reliance.

Each of the pertinent controls for the other three objectives (invoices are accurately prepared and properly recorded, entries to accounts receivable are authorized, and cash receipts are adequately controlled and accurately recorded) should be analyzed in a similar manner. For each source of error or irregularity, the auditor identifies those pertinent procedures that prevent or detect the error. He then decides which of the pertinent procedures are to be tested statistically and designs the test so that the set of procedures can be evaluated as one procedure at a specified risk level.

Risk of Overauditing. Compliance tests may also be a source of risk of overauditing. When the auditor decides that compliance with some pertinent accounting control procedure may not be satisfactory, there must be a change in the substantive audit program. These changes affect either the nature, extent, or timing of the planned audit procedures. Thus when the statistical evidence incorrectly leads the auditor to make these changes, unnecessary auditing expense is incurred. Planning the compliance tests allows the auditor the opportunity to control the risk of unnecessary expense. This risk equals the probability that the compliance tests will result in indicating lack of satisfactory compliance when, in fact, compliance is good.

The auditor can control the allowable risk by specifying the probability that the achieved upper precision limit should exceed the threshold rate (P_0) when in fact, the rate of compliance deviation does not exceed P_1 as explained in chapter 4. Considering the appropriate level for this risk involves balancing the cost of increasing the sample size necessary to lower the risk for compliance purposes with the expected cost reduction from avoiding changing the planned audit program for the substantive procedures.

For example, if the auditor designates $P_0 = .05$ and $P_1 = .01$ with the sampling risk specified at .05 (one-sided reliability equal to .95), using a sample of 60 for compliance purposes entails a risk of overauditing equal to .46, while increasing the sample to 100 reduces the risk to .26. To calculate whether the increased number of observations is worthwhile, the auditor needs to estimate the cost of the 40 additional observations and compare that with the expected increase in auditing costs associated with the expanded audit program. The expected increase equals the difference in risk .20 (.46 − .26) multiplied by the cost of the additional audit work when compliance is found not to be satisfactory. If it is found that the cost of the 40 additional observations is lower, then the auditor might continue to add observations until the additional cost equals the reduction in expected auditing costs.

Sample Design. Deciding the risk levels for testing compliance with each pertinent accounting control precedes the detailed design of the sample. The sample design consists of (1) defining the attribute to be tested, (2) specifying the sampling unit and frame, (3) determining the appropriate sample size, and (4) specifying the selection method. While the test of each control is planned separately, the field work is designed to accomplish several compliance tests with the same sample. Moreover, combining the compliance tests with the substantive tests of the transactions (known as a dual-purpose test) must be considered.

Defining the attribute—A careful definition of each procedural deviation to be tested is required. Defining a procedural deviation entails considering the actions to be taken if compliance is determined to be unsatisfactory. For example, in testing purchases, the auditor determines that the pertinent control against unauthorized purchases is requiring all purchases to be properly authorized. Proper authorization is different for different classes of purchases—some (such as purchase of utilities) require only general authorization, while some require specific authorization in the form of signed purchase orders. Using a broad definition of the procedural deviation as "purchase not properly authorized" would apply to every purchase and would be appropriate only if the auditor determines that he could suitably alter his program knowing only that too many purchases were made without evidence of proper authorization. It seems, however, that this broad definition would usually be unsuitable—why extend the substantive procedures for all accounts affected by purchases on the basis of the failure of a few purchases to be supported by a signed purchase order?

An alternative is to define the procedural deviation narrowly—purchase *orders* not properly authorized. Using this definition, the evidence of unsatisfactory compliance would only force the auditor to alter substantive tests of transactions or balances that should be covered by purchase orders. Some auditors might even subdivide this control further, by regarding purchases of raw material and supplies employed in the manufacture of products separately from other purchases that are not evidenced by receiving records. Again, the appropriate definition is determined by what actions the auditor would take in the event compliance is found to be unsatisfactory. Remember that the auditor need test only those pertinent controls on which he plans to rely in limiting his substantive tests. Consequently, the auditor has the option of not relying on a pertinent control if it is more efficient to eliminate the compliance test and instead increase substantive testing.

The other major requirement in defining the compliance deviation is clarity. It is imperative that the auditor who is conducting the test understand exactly what constitutes a compliance deviation. For example, if a signature is required for authorization of a purchase and none exists, but the auditor is told that verbal approval was given, a compliance deviation has occurred. This means that the details of "properly authorized" must be spelled out in advance.

When the compliance test pertains to a set of procedures, the auditor needs to define the attribute for each procedure and, in addition, state what constitutes an occurrence. If a prevention control and detection control are both involved, an occurrence corresponds to lack of compliance with both procedures on the same transaction. When compliance with the set of controls requires compliance with each, an occurrence corresponds to a lack of compliance with any one of the procedures.

Sampling unit and frame—Once the procedural deviations have been clearly and usefully defined, the auditor specifies an appropriate sampling unit and corresponding frame. The only requirements of the frame are that it be complete and allow the auditor to perform the audit procedures. Beyond these requirements, the auditor selects the frame that appears to be most convenient in terms of time and cost.

Since tests of compliance are often combined with substantive tests of transactions (dual-purpose tests), the requirement that the frame allow the desired audit procedures is very important. This can be illustrated by considering a test of purchase transactions. The auditor finds the following available frames: (1) purchase order file, (2) invoice register, and (3) cash disbursements book or check register. When substantive tests of purchases are involved, the fact that the cash disbursement book or the check register can only be used to test assertions relating to payments made but not to payments due becomes important in selecting the frame.

This example also illustrates the requirements of completeness and convenience. The purchase order file might be considered incomplete (depending on the definition of the population), since some routine purchases such as utility services do not require a purchase order. Comparing the invoice register and the check register for convenience, the auditor must consider that using the check register and defining the corresponding sampling unit as a check may not be as convenient as using the invoice register with the individual invoice as the sampling unit. Unless some form of multistage sampling can be employed, using the check as the sampling unit entails applying all auditing procedures to each check as well as to all the invoices paid by the selected check. Obviously, this requires much more time and effort than applying the same set of procedures to a single selected invoice.

Determining sample size—Determining sample sizes for compliance tests (or dual-purpose tests) can be done quite easily using tables or a time sharing computer program.[14] Most of the available tables for determining a fixed sample size do not provide control over the risk of overauditing. When controlling this risk is desirable, a computer program is necessary.

The auditor can select either a fixed sample size or a sequential plan as described in chapter 4. The use of the sequential plan should be considered, especially when the anticipated occurrence rate is either very small or quite large. Expected sample sizes under these conditions will normally be smaller than the corresponding fixed-size sample. On the other hand, if the population occurrence rate actually is of moderate size, the ultimate sample size under the sequential plan may exceed the corresponding fixed-size sample.

Since most samples are used for testing several accounting controls, it is necessary to determine a sample size, or sequential plan, for each pertinent control and then use the largest sample or the sequential plan with the largest initial and incremental sizes. If it is convenient, the auditor may stop examining selected sampling units for compliance with specific controls when the sample size required for that control has been met. This presumes, however, that the units are examined in the random order of selection.

Selecting the samples—The basic choice is between using an unrestricted random sample and a pps sample. The latter type of sampling offers the distinct advantage of relating the incidence of compliance deviation directly to dollars by expressing the fraction of the dollars that are associated with a compliance deviation.

Since the attribute tables that are often employed in compliance tests are based on unrestricted random sampling, the auditor should avoid using systematic sampling whenever possible. When circumstances are such that systematic selection is the only practical method, several random starts should be used to help guard against unknown patterns within the frame.

---

14. See chapter 9 for a description of the computer programs.

Planned Reliance

Having designed the extent of substantive tests as well as the compliance tests for selected degrees of reliance, the auditor can adopt the degree of reliance that represents the least costly combination. This step completes the preliminary design of the audit program.

Final Design of Compliance Tests

The preliminary design proceeds by audit area, and within each audit area, the compliance tests are tentatively planned to meet the auditor's planned degree of reliance. In some cases, pertinent controls for one area also serve as pertinent controls for another. For example, controls over cash receipts affect both accounts receivable and cash, and controls over shipments affect sales, accounts receivable, and inventory. In other cases, substantive tests of transactions may involve examining the same transactions as one or more of the tests of compliance.

In these circumstances, the auditor needs to plan the field work in a way to minimize the total audit effort. This means combining into a single sample all those audit procedures that may have initially been planned separately. When some of these are substantive procedures, the tests become dual-purpose tests. An example of this is included in the following chapter.

Execution and Evaluation

Careful execution of the planned statistical tests and appropriate evaluation of the resulting sample evidence are required in order to realize the benefits of the planning process. A major concern during execution is making sure that the planned audit procedures are, in fact, accomplished. Control of these so-called nonsampling errors is vital, because all the statistical projections depend upon the correct evaluation of each sampling unit. Control can be exercised by making sure that persons doing the field work understand exactly what they are to do, that an "audited amount" or "sample occurrence" is unambiguously defined, that calculations are checked for accuracy, that selected items are examined without substitutions, and that all steps are properly documented.

One common problem is what to do about missing evidence concerning a selected sampling unit. Suppose, for example, that a confirmation request is returned by the post office as undeliverable, or a positive confirmation request is not returned even after sending a second request, or that a selected bill of lading cannot be located. Such occurrences require some action; they cannot simply be ignored.

The appropriate action depends upon the circumstances. In the case of a non-response to a confirmation request, there may be an opportunity to perform alternative procedures that will satisfactorily determine whether the account exists and is recorded accurately. When this is not possible, the auditor might tentatively evaluate a non-response as if it were totally in error. When this tentative evaluation does not indicate a potential material error, nothing more is required for a statistical conclusion.

Evaluating a non-response or missing document as if the sampling unit were totally in error is sound conservative practice, but the auditor must not stop there. There is an ever-present danger that far too much attention may be focussed on the mechanics of statistical sampling and that, consequently, the auditor may fail to exercise sound audit judgment. The case of undeliverable confirmation requests illustrates this. The auditor should investigate and try to determine the cause in addition to making the statistical evaluation. This problem is discussed further in chapter 10.

Evaluating Compliance Tests

The results of compliance tests are evaluated according to the plan to determine whether the planned reliance is justified. For a fixed-size sample, this entails determining the achieved upper precision limit at the specified one-sided reliability (R_1) for each of the pertinent controls or set of pertinent controls tested with the sample. The achieved upper precision limit either does or does not exceed the specified threshold rate (P_0). If it does not, the auditor may rely on the control as planned. His risk in doing this equals the complement of the one-sided reliability ($1 - R_1$).

When the achieved upper precision limit does exceed P_0, the statistical evidence fails to corroborate the auditor's planned reliance. Consequently, some action must be taken. The appropriate action must be determined from the circumstances, but in general will involve changing the timing, nature, or extent of some of the planned substantive tests. The particular substantive procedures involved depend on the type of error that the control or set of controls was designed to prevent. As an illustration, if the auditor finds that the number of sales made without checking credit could exceed the threshold rate, the scope of the audit program might be expanded with respect to collectibility of accounts receivable.

The same decisions apply when the auditor uses a pps sample and determines the proportion of the total transactions associated with a particular compliance deviation. In the previous example involving sales, the auditor would determine the upper precision limit of the proportion of sales dollars (instead of number of sales) associated with not checking the customer's credit. The planned substantive tests of collectibility would be extended whenever this proportion exceeded the established monetary limit.

Similar decisions are appropriate when the auditor uses a sequential sample. In contrast to the fixed-size sample, a sequential sample does not automatically provide a point estimate of the extent of compliance deviations, but does allow the auditor to decide whether compliance is or is not satisfactory for the planned reliance.

The evaluation of the compliance portions of dual-purpose tests follows this same general scheme. Evidence that fails to corroborate the planned reliance calls for extending the substantive procedures. This may entail extending the substantive portion of the test on a sample basis, or performing additional substantive tests. If the tentative audit program called for examining transactions for only part of the year, consideration should be given to extending the population of transactions to the entire year for the extended substantive tests.

Evaluating Substantive Tests

Evaluation of the statistical substantive tests entails using a valid evaluation technique to determine whether the recorded amount could be materially misstated. As previously described, this can be done using either a positive or negative approach. The negative approach is described here.

Using the negative approach, the auditor computes either an upper precision limit or an upper bound for the monetary error from all independent statistical tests of details pertaining to a particular recorded amount. If this upper precision limit or upper bound exceeds the amount (M)—the material amount allocated to the statistical tests of details—the auditor regards the statistical evidence as indicating the possibility of material error. This calls for some action to determine whether there is indeed material misstatement and, if so, what adjustment would be appropriate in the circumstances. While the specific circumstances determine the most appropriate action, one alternative source of additional information would be a precise statistical estimate of the amount of monetary error.

The standards for supporting an accounting adjustment may be quite different from those required to decide whether the recorded amount could be materially misstated. If the auditor concludes that a statistical sample is the most appropriate source of evidence for the purpose of estimating the adjustment to be recorded, a precision and reliability must be selected that is proper for supporting an accounting entry. As a guideline, using an interval reliability of at least .95 and a precision no larger than $.5M$ would appear reasonable in this circumstance. Using this guideline allows the auditor to assert that there is a confidence of .95, that the precision interval contains the true amount, and, furthermore, that whatever the true amount, it would not materially differ from the adjusted amount.

When the achieved upper precision limit or upper bound of monetary error does not exceed the amount (M), the auditor decides that there is no material misstatement. The sampling risk of being incorrect in this conclusion equals the complement of the one-sided reliability $(1.0 - R_1)$ or $(1.0 - R)/2$ if an interval reliability (R) was used.

The statistical techniques used to calculate the achieved upper precision limit, or upper bound, depend upon the sample design and the observed results. First, consider the stratified design in which the recorded amounts were used for the basis of stratification as well as for determining the appropriate sample size. The sample size was determined to meet a desired precision and reliability using the stratified mean estimator based on the recorded amounts.

If no differences are observed, the sample estimate of the monetary error is zero. The statistical difficulty is twofold: (1) obtaining a usable estimate of the standard error of this estimate and (2) being sure that the sampling distribution is approximately normal. The achieved standard error of the mean estimator is a usable upper bound for the standard error of the difference estimator, but a concern in using this is whether the sampling distribution is approximately normal. It may not be.

An alternative that does not assume normality was described in chapter 6. This alternative involved the following procedures:

1. Determine the achieved upper precision limit for the proportion of sampling units that could contain monetary error at a desired reliability (R_1).
2. Multiply the number of population units within each stratum times the largest recorded amount in each stratum and add the products together.
3. Multiply the achieved upper precision limit times the sum of the products.

This procedure results in an upper bound on the amount of overstatement in the population. It can be used whenever the sample items have been allocated to the strata in proportion to the number of population units within the stratum multiplied times the largest recorded amount in the stratum.

If only a few differences are observed, the auditor may also use the bound just described or compute the sample estimate of the monetary error using, say, the difference estimator. As described in chapter 6, the achieved precision of the mean estimator may be used to obtain a bound on the overstatement error, provided the square of the stratum coefficient of variation of recorded amounts exceeds the proportion of non-zero differences within the stratum.

If several differences are observed and the auditor is reasonably satisfied that they are representative of the magnitudes of errors to be expected in the specific situation, he can use the approximate method described in chapters 5 and 6 for computing the standard error associated with any one of the estimators—difference, ratio, or regression.

If many differences are observed, the sample estimate of the monetary error can be determined using the difference, ratio, or regression estimator. The estimated standard error is determined from the sample data in accordance with the formulas presented in chapter 5 or 6. Using the negative approach, there is no need to make any adjustments when the achieved precision differs from the planned precision. As described in chapter 3, the risk of overauditing may be affected, but the sampling risk remains at the planned level. When the auditor uses a positive approach, an adjustment is required if the sampling risk is to be maintained at the planned level. This was described in chapter 3.

A pps sample can also be evaluated in different ways that depend upon the number of differences observed. When no differences are observed, an upper bound for the monetary error is computed by first determining an upper precision limit for the fraction of the recorded amount that might be in error and extending this times the total recorded amount. This yields a conservative bound on the potential error in the unobserved portion of the population.

For example, a pps sample of 300 units that has no observed differences between audited and recorded amounts permits concluding that, at most, 1 percent of the total recorded amount could be in error at a one-sided reliability of .95. If the total recorded amount is $2 million, the achieved upper bound for monetary error is therefore $20,000 ($2,000,000 × .01).

When some differences are observed, the auditor can select one of the methods presented in chapter 6 for determining an upper bound when only a few differences are observed.

When many differences are observed, the standard pps estimator can be used together with the estimated standard error.

Conclusion

Integrating statistical sampling into auditing requires careful planning, proper execution, and appropriate evaluation of the sample results. Each of these subjects has been discussed in this chapter. The objective throughout has been to describe methods that can be used by the auditor to achieve overall control of sampling risk as well as exercise some controls over auditing costs.

How practical are these methods? Much of what is described here is currently being done in practice, but not always in the manner described. Some of the suggestions call for doing something different from what is currently being done. The particular suggestions, however, are not so important as following some consistent procedure for planning, executing, and evaluating sample results. In the next chapter an example of a particular situation is discussed, and the methods described here are applied.

8

Case Study: The Maxwell Manufacturing Company

The principal objective of this chapter is to illustrate how statistical sampling can be integrated into the audit planning process. The framework for integration was described in chapter 7. Using this framework, the sales and accounts receivable portion of the audit for Maxwell Manufacturing Company is described. The many interrelationships that exist between this area of the audit and other areas such as inventory and cash are not described in detail.

Following the audit process flowchart shown in chapter 7, the discussion begins with a detailed description of the preliminary design of the audit program and follows the subsequent implementation and change of that program. As previously noted, the preliminary design results only in a tentative audit program based on the auditor's current information. As the work progresses, more information becomes available and the preliminary program is modified accordingly. Thus, the final audit program evolves through the process of planning, executing the tests, evaluating the resulting evidence, and revising the plan.

The description of the decisions made for this case serve only as an illustration of the types of decisions the auditor would make in using statistical sampling. Other auditors might well make decisions that differ from those cited here.

Maxwell Manufacturing Company is a medium-sized firm which manufactures hand tools such as hammers and screwdrivers. Its products are distributed nationally through commission-paid representatives who sell building materials, power tools, and other items to hardware stores and building supply outlets. The firm's products are generally displayed on racks by the retailers.

Maxwell has experienced a growth rate of approximately 10 percent over the past several years. The client currently has approximately 6,000 regular customers. These customers average about eight orders per year, and the average order size is about \$200. Thus, sales amounted to about \$10 million last year. Accounts receivable comprise nearly 34 percent of the total assets.

The auditor anticipated that the balance sheet for the current year would closely resemble the following:

Cash	$ 500,000	Accounts payable	$ 700,000
Accounts receivable	1,400,000	Accrued taxes	100,000
Inventory	1,200,000	Bank debt	1,500,000
Plant and equipment	1,000,000	Equity	1,800,000
	$4,100,000		$4,100,000

Furthermore, the auditor anticipated that profit would amount to about $2 million before taxes on the $10 million in sales.

The company has a September 30 year end. This is the second year for the current auditors. Two years ago the company installed an IBM 360/30 computer. During the first year's audit, the computer system was thoroughly reviewed—both general controls and applications controls. An internal control questionnaire was used as part of the evaluation procedure, and flowcharts of each transaction cycle were prepared.

During the current year, the auditors reviewed and updated the questionnaire and determined that the flowcharts previously prepared were still applicable.

Using the information gained from these procedures, together with his preliminary evaluation of the accounting controls, the tentative audit program was designed. Exhibits 8-1 through 8-4 on pages 158–165 are the flowcharts and descriptions pertaining to the sales/account receivable area.

The auditor's preliminary evaluation of the system included the observation that hash totals of customer order numbers were prepared after processing, and, consequently, any loss of documents prior to keypunching would be undetected. As a compensating control, it was noted that there is a numerical control over the open order file and orders open for unusual periods of time are followed up.

In other respects, the auditor thought the system was quite good. Exhibit 8-5 (pages 166–167), the interdependence matrix, summarizes his findings by matching the key elements of control against the major sources of error.[1] In each case, except for the above-noted weakness, the auditor decided that the system was strong enough to justify a high degree of reliance provided there was satisfactory compliance.

Preliminary Design

The case begins with the description of the preliminary design process following the eight steps listed in chapter 7. The auditor's information at this point in time consists of preliminary evaluation of the effectiveness of internal accounting control procedures and the results of the diagnostic analysis.

Audit Objectives

The first logical step in designing the preliminary audit program for sales and receivables is to identify the auditor's general objectives. The basic objective concerning accounts receivable is to determine whether they are fairly presented in

1. The form for this matrix was suggested by James K. Loebbecke.

accordance with generally accepted accounting principles applied on a consistent basis. To achieve this general objective, the auditor adopts the following specific objectives:

1. Determine whether accounts receivable are materially overstated with respect to existence, recorded amount, collectibility, and period cutoffs.
2. Determine whether accounts receivable are materially understated.
3. Determine whether accounts receivable are properly classified and whether there is adequate disclosure with respect to generally accepted accounting principles.

The related objectives with respect to sales are these:

4. Determine whether sales are materially overstated.
5. Determine whether sales are materially understated.

The auditor may subdivide any of these objectives into a number of subsidiary objectives. For example, the first objective may be divided as follows:

1A. Determine whether accounts receivable are materially overstated with respect to existence and recorded amounts.

1B. Determine whether accounts receivable are materially overstated with respect to collectibility.

1C. Determine whether accounts receivable are materially overstated with respect to period cutoffs.

The auditor needs evidence to support each of these objectives. Because of the interrelationships between the accounts, this does not mean that each objective requires its own independent set of evidence. In fact, by examining the types of available evidence supporting each objective, it will be apparent that a test frequently provides evidence for more than one objective.

Materiality

The auditor decided that $50,000 would be an appropriate material amount for sales and accounts receivable. This amount is determined judgmentally after considering the size of the accounts receivable balance in relation to the total assets, the expected income, and the outstanding debt. The auditor also made allowance for allowable monetary errors in other parts of the audit.

Identifying the Tentative Set of Substantive Auditing Procedures

The next logical step is to identify those procedures that are to be used to obtain evidence pertinent to each of the audit objectives. The procedures selected by the auditor at this stage of the program design reflect his current knowledge. As he learns more, he may revise the set of procedures that he uses.

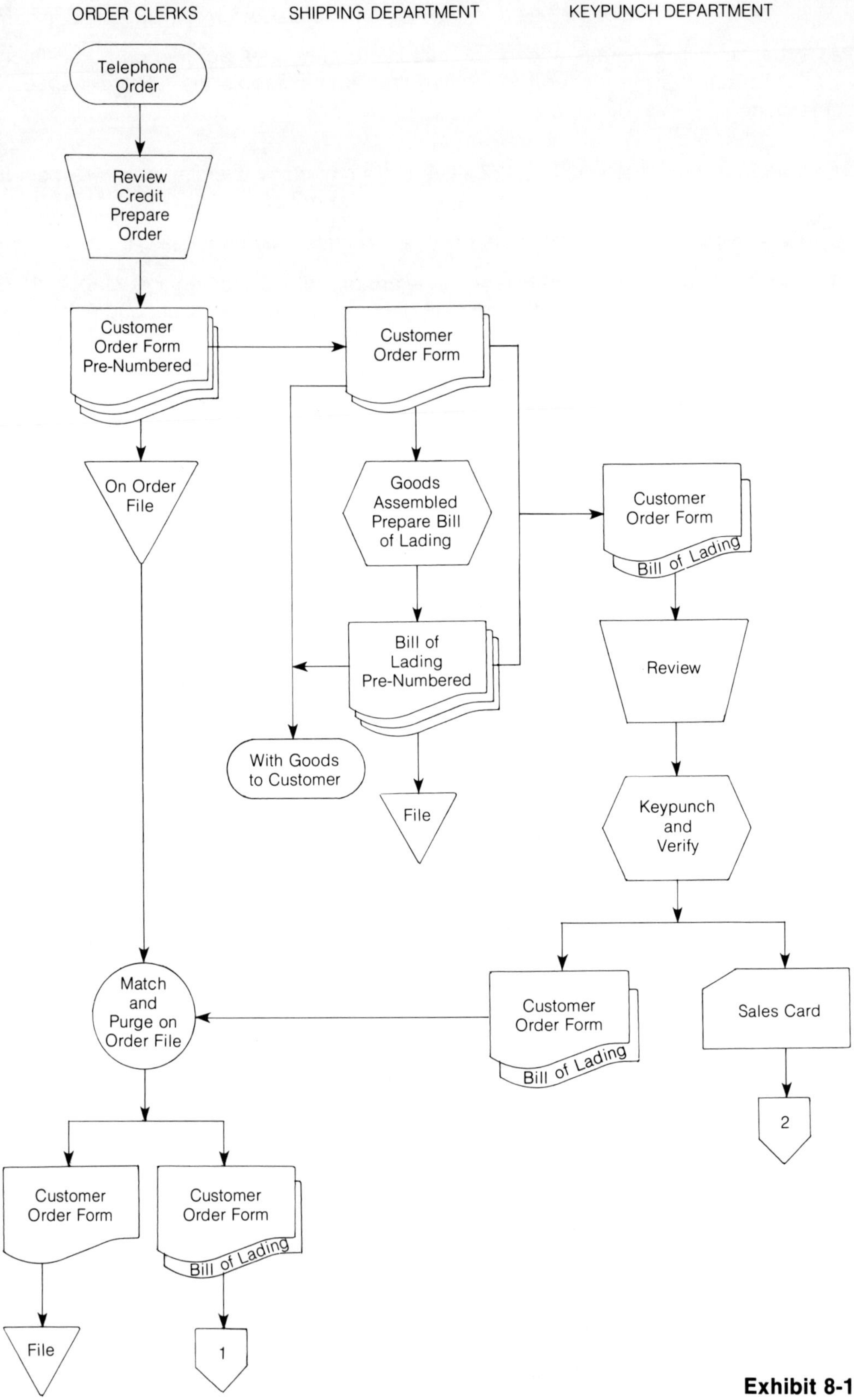
ORDER CLERKS
SHIPPING DEPARTMENT
KEYPUNCH DEPARTMENT
Telephone Order
Review Credit Prepare Order
Customer Order Form Pre-Numbered
On Order File
Customer Order Form
Goods Assembled Prepare Bill of Lading
Bill of Lading Pre-Numbered
With Goods to Customer
File
Customer Order Form
Bill of Lading
Review
Keypunch and Verify
Customer Order Form
Bill of Lading
Sales Card
2
Match and Purge on Order File
Customer Order Form
Customer Order Form
Bill of Lading
File
1

Exhibit 8-1

Order clerk obtains the customer number from a master customer list. Ascertains whether the customer has an active account, has not exceeded his credit limit, and has no balance more than 30 days past due. Initials a copy of the customer order form to indicate this procedure was followed. Sales to new customers and to recently inactive customers are approved by Mr. Jones (V.P., Sales). All orders exceeding $500 require credit approval from Mr. Gordon (Controller). These approvals are indicated by a signature on the customer order form.

Shipping clerk manually changes the two customer order forms to record under-shipments. First copy of customer order form and second copy of bill of lading are sent to keypunching. The order number is entered on the third copy of the bill of lading which is filed by bill-of-lading number in the shipping department.

The keypunching department batches the customer order forms together with the bills of lading which represent the prior day's shipments. A sales card is prepared containing the customer order form number, salesman's number, customer number, part number, quantity shipped, and date shipped. The customer number, part number, and quantity shipped fields are verified by a second keypunch operator. No hash total of the customer order number is prepared prior to keypunching to prevent loss of documents.

The order clerks compare the bill-of-lading quantities to quantities shipped per the customer order form, match the first copy of the customer order form to the second copy in the on-order file, and purge the file. The clerk initials the file copy of the customer order form to indicate that these procedures were completed.

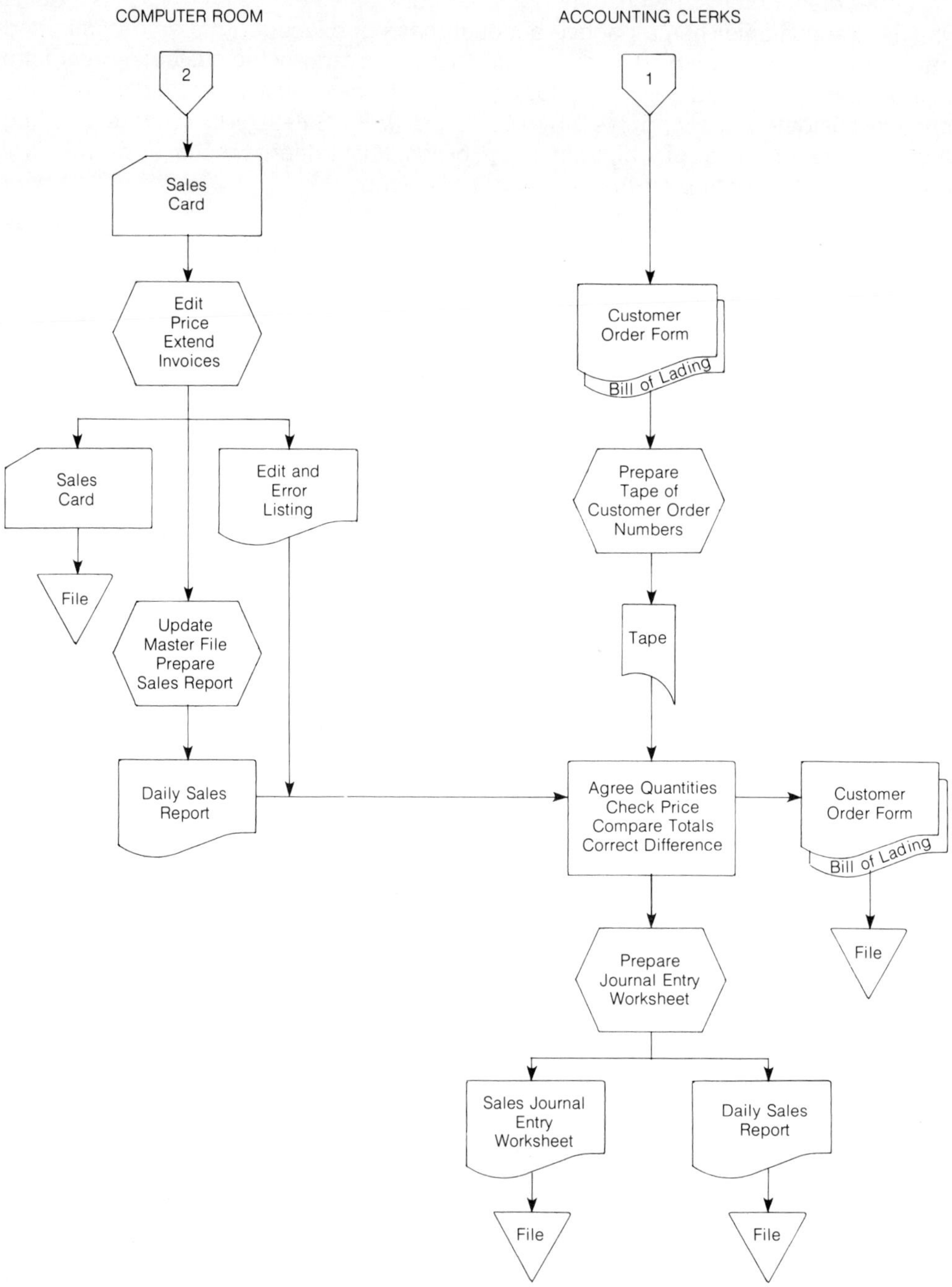
COMPUTER ROOM
ACCOUNTING CLERKS
2
1
Sales Card
Edit Price Extend Invoices
Sales Card
File
Edit and Error Listing
Update Master File Prepare Sales Report
Daily Sales Report
Customer Order Form
Bill of Lading
Prepare Tape of Customer Order Numbers
Tape
Agree Quantities Check Price Compare Totals Correct Difference
Customer Order Form
Bill of Lading
File
Prepare Journal Entry Worksheet
Sales Journal Entry Worksheet
Daily Sales Report
File
File

Exhibit 8-2

Sales cards are processed daily. Sales cards are used as input to a pricing program that prices the sales from the selling price master file. The program rejects transactions for invalid data or unmatched part numbers and lists them on an edit and error listing. A daily sales report is prepared listing the details of each sale in customer sequence. This report includes totals on the amount of sales and sales tax (if any) and a hash total of customer order form numbers.

The daily sales report is used to update the accounts receivable master file. Details of transactions having customer numbers not appearing on the master file are listed on an error listing. The error listing including control totals is sent to the accounting clerks with a copy retained by the supervisor. The supervisor uses the listing to insure that all rejected transactions are corrected and posted to the receivable master files.

Accounting clerks check agreement of quantities on the order form, bill of lading, and daily sales report and initial the customer order form. Prices are compared to a master price list on a test basis and the customer order form is initialed. All orders over $500 are reviewed for necessary approvals. An adding machine tape of customer order form numbers is prepared. This is compared to the hash total on the daily sales report. Errors on the error listing and discrepancies in the hash totals are investigated and adjusted sales cards are prepared and processed the following day. The daily totals of sales and sales tax are posted to a monthly sales worksheet, which is used to prepare the monthly sales journal entry. The order forms are filed numerically, and each week open orders, indicated by gaps in the file, are followed up.

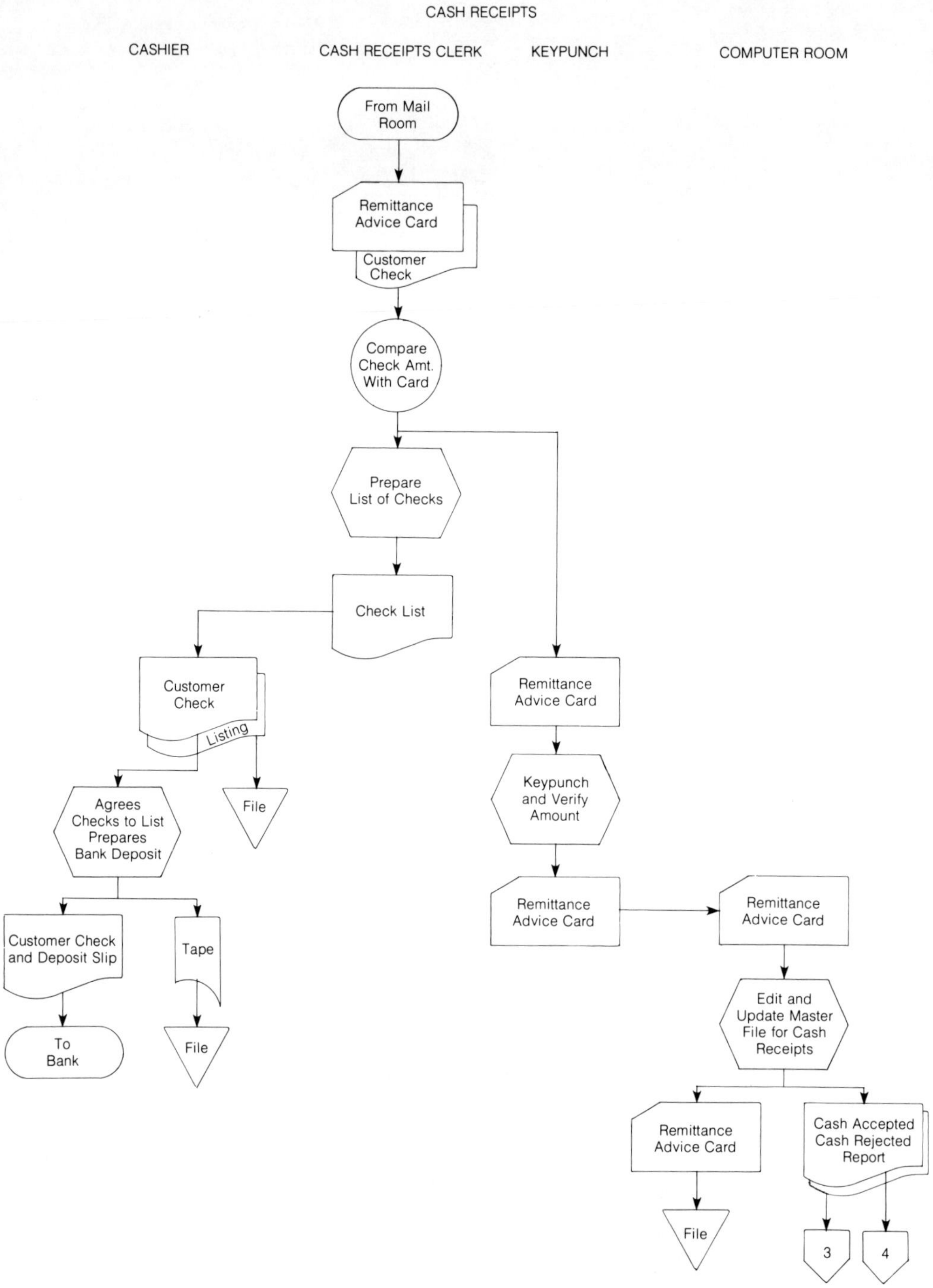
CASH RECEIPTS
CASHIER
CASH RECEIPTS CLERK
KEYPUNCH
COMPUTER ROOM
From Mail Room
Remittance Advice Card
Customer Check
Compare Check Amt. With Card
Prepare List of Checks
Check List
Customer Check
Listing
File
Agrees Checks to List Prepares Bank Deposit
Customer Check and Deposit Slip
Tape
To Bank
File
Remittance Advice Card
Keypunch and Verify Amount
Remittance Advice Card
Remittance Advice Card
Edit and Update Master File for Cash Receipts
Remittance Advice Card
Cash Accepted Cash Rejected Report
File
3
4

Exhibit 8-3

Mail addressed to the collection department is forwarded to the cash receipts clerks, who prepare a list of the checks received and restrictively endorse the checks. The clerks compare the amount of the check with the amount manually entered by the customer on the prepunched remittance advice card and initial the check. The remittance advice card is mailed with the monthly statement, and if it is not returned, the clerks enter the data on a form for keypunching.

The cashier compares the checks to the list, initials the list, and prepares the deposit. Deposits are made daily, and the authenticated deposit slip is returned to Mr. Gordon (Controller).

Miss Henderson (keypunch supervisor) reviews the remittance advice cards and forms and gives them to the operators who keypunch the receipt date and amount or complete a new card, as appropriate. Date and amount are key verified.

The program to update the A/R master file offsets the receipt against all orders included in the statement if receipt equals the statement amount. Otherwise the remittance is entered as an unallocated credit to the customer's account. Unallocated credits are posted to a suspense account until identified with specific orders. Unapplied receipts or credits are listed separately on the cash accepted-cash rejected report. This report also contains transactions that were applied to open orders and those rejected because a customer number could not be located. Also shows control totals for the master file including numbers of customers, total open receivables, and total unallocated receipts.

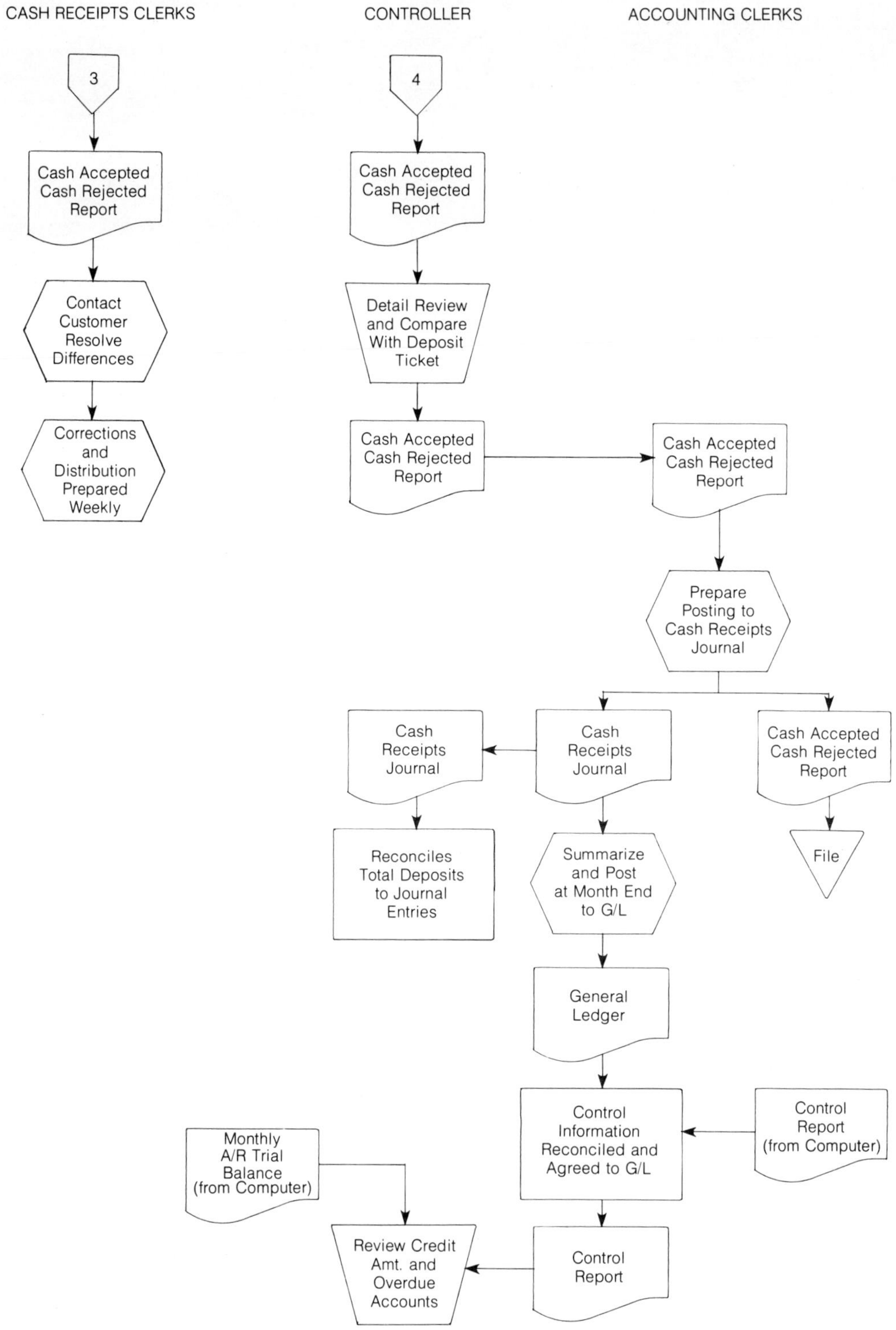

CASH RECEIPTS CLERKS
CONTROLLER
ACCOUNTING CLERKS
3
Cash Accepted Cash Rejected Report
Contact Customer Resolve Differences
Corrections and Distribution Prepared Weekly
4
Cash Accepted Cash Rejected Report
Detail Review and Compare With Deposit Ticket
Cash Accepted Cash Rejected Report
Cash Accepted Cash Rejected Report
Prepare Posting to Cash Receipts Journal
Cash Receipts Journal
Cash Receipts Journal
Cash Accepted Cash Rejected Report
Reconciles Total Deposits to Journal Entries
Summarize and Post at Month End to G/L
File
General Ledger
Control Information Reconciled and Agreed to G/L
Control Report (from Computer)
Monthly A/R Trial Balance (from Computer)
Review Credit Amt. and Overdue Accounts
Control Report

Exhibit 8-4

Cash receipt clerks follow up rejected items on the report. Customers are contacted when necessary to ascertain proper distribution of unallocated amounts. The corrections and distributions are batched weekly and sent via keypunching to the computer room for processing against the master file and suspense account. A weekly suspense account update report is produced and follows the same flow as the cash accepted-cash rejected report.

At month end, Mr. Gordon reconciles total deposits per the bank statement and deposit slips to the cash receipts journal entries. Mr. Gordon also investigates and controls all chargebacks from the bank.

When monthly statements are prepared the last page of the run contains the following control information: the total amount of open orders on the A/R master file, the total of unapplied credits in the suspense account and the total amount due per the monthly customer statements. The statements, remittance advice cards, and control information are sent to the accounting clerks where the control information is reconciled and agreed to the general ledger.

After the monthly statements are run, a monthly aging report of accounts receivable is prepared and sent to Mr. Gordon. The aging includes the total unallocated payment and the credit limit. Mr. Gordon compares the balance on this aging to the control totals generated with the customer statements. He reviews the individual accounts and compares the total of each account with the credit limit.

Exhibit 8-5

Transaction Error Types	Identified[1] Controls	Maximum Possible Reliance	Weaknesses	Effect on Financial Statements[2]: Cash	A/R	Allow D/Acct.	Inv.	Net Sales	Costs Expenses	Compliance Tests: Procedures[3]	Planned Reliance	Substantive Tests: Procedures[3]	Planned Risk
SALES													
Made to unacceptable credit risk	A2 E1 E2 E3 E4	high				u			u	T1 T2	moderate	P4 P5 P7	moderate
Shipment unauthorized	A1 A2 B1 E1	high		no effect						T1 T2	none	—	
Shipment inaccurate	A3	high			o/u		o/u	o/u	o/u	T1 T3	moderate	T18	moderate
Invoice not prepared	A1 B1	low	inadequate control over keypunch input		u			u		T1	none	T17 T18 T19 T6 T7 P6 P7	low
Invoice inaccurate	A3 B2 B3 B4 D2	high			o/u			o/u		T1 T3 T4	moderate	P1 P2 P9 P10	moderate
Invoice improperly recorded	B3 B4 D1 D2	high			o/u			o/u		T1	moderate	T20 T22 T24 T25 P1 P2 P7 P9 P10	moderate
CASH RECEIPTS													
Cash receipts not recorded	C1 C2 C3 C4 C5 D1	high		u	o					T1 T9	moderate	P1 P2 P6 T8 P8	moderate
Amount recorded incorrectly	D2	high		o/u	o/u					T11 T15	moderate	T10 T14 T15 T21 T22 T24 T25 P1 P2 P6	moderate
OTHER													
Journal entries not authorized	D1 D3 D4 D5	moderate		o/u	o/u	o/u		o/u		T1 T12	moderate	T13 T23 T24 T25 P3 P9	moderate
Uncollectable accounts not written off	E4 E5	high			o	o				T1	low	T24 T25 P4 P5	low
Improper classification or inadequate disclosure											none	P3	low

1. *Legend of Identified Controls*

A1 Sales orders prenumbered and controlled
A2 Sales orders independently authorized or approved
A3 Quantities ordered, shipped, recorded are independently checked
B1 Sales order (see A1 and A2) is shipping order
B2 Price list maintained (computerized)
B3 Daily sales independently received
B4 Sales detail independently reconciled to accounts receivable
C1 Prelisting of cash receipts
C2 Remittance advice used and compared to checks received
C3 Deposit slip independently checked to cash receipts journal
C4 Receipts deposited intact daily
C5 Bank account reconciled monthly by independent person
D1 Detailed accounts receivable reconciled to control monthly
D2 Monthly statements sent to customer
D3 Credit memos require approval
D4 Returned goods require receiver
D5 Advances to employees require authorization
E1 New customers require separate approval
E2 Sales orders matched against delinquent list and credit limits before processing
E3 Credit limits periodically reviewed
E4 Monthly aged trial balance prepared and delinquent accounts reviewed
E5 Bad debt write-offs require approval

2. o Overstatement error
u Understatement error
o/u Over or understatement error

3. *Legend of Tests and Procedures*

T1 Observation and/or inquiry

For customer orders:
T2 Ascertain approval or authorization
T3 Ascertain initialing of customer order signifying agreeing quantities on orders, bills of lading, and daily sales report; check whether quantities agree
T4 Ascertain initialing of customer order signifying that price was compared to price list; check whether prices agree
T5 Trace to monthly statement, A/R trial balance and subsequent credit (if any)
T6 Scan file for gaps in numbers
T7 Review on-order file for evidence of orders shipped but not billed

For those paid:
T8 Trace deposit slip to bank statement
T9 Trace to listing of remittances and ascertain whether cash receipts clerk agreed check amount to remittance advice
T10 Determine credit to proper customer
T11 Examine noncash credits for proper approval

For credit memos:
T12 Ascertain proper support and authorization
T13 Verify prices and extensions

For suspense report items:
T14 Trace to cash accept/reject report
T15 Examine support
T16 Trace to application on A/R trial balance

For bills of lading:
T17 Trace to sales order in accounting
T18 Agree quantities to sales order
T19 Trace sales order to daily report; note delay from shipping date

Test footings and postings:
T20 Daily sales report
T21 Cash accept/reject report
T22 Weekly suspense report
T23 Large, unusual journal entries
T24 Monthly statement run
T25 Aged trial balance

For accounts receivable:
P1 Request confirmation of the recorded amounts
P2 Analyze and test the account from the date of confirmation to the closing date
P3 Review classification and disclosure
P4 Test aging
P5 Test subsequent collections
P6 Test period cutoffs of sales and inventory
P7 Analyze and evaluate allowance for doubtful accounts
P8 Reconcile accounts receivable detail file to general ledger balances
P9 Reconcile total credit to sales to debit to accounts receivable
P10 Reconcile sales order volume to capabilities, capacities, etc.

The question whether a particular test is a substantive test or a compliance test has been resolved here by reference to its major purpose. Nearly every test the auditor performs has both substantive and compliance aspects. Nevertheless, when the principal objective is to determine, on the basis of documented evidence, whether an accounting control is operating as planned, the test will be regarded as a test of compliance, and when the principal objective is to determine whether the dollar amounts are correct, the test will be regarded as substantive. Tests having both these objectives will be regarded as dual-purpose tests.

A test that might be regarded as a compliance test when a system is judged to be strong may be regarded as a substantive test when the system is weak. For Maxwell, the apparent weakness in a basic input control (lack of batch control totals over input data) resulted in the test of the bills of lading being regarded primarily as a substantive test to determine whether all sales were properly recorded.

Likewise any test designed primarily to test compliance with an apparently strong accounting control may be replaced by a substantive test when the evidence indicates that compliance may be unsatisfactory.

The list of the tentative procedures appears on page 167 immediately following the interdependence matrix (exhibit 8-5). Those procedures that are done as part of the tests of transactions are denoted by a "T." These procedures include compliance tests, substantive tests, and dual-purpose tests. Procedures that are direct tests of the accounts receivable balance are denoted by "P."

These procedures are shown in the interdependence matrix where each is classified by the type of transaction error it can detect and according to whether it is a compliance or substantive procedure. Only those procedures that examine documented evidence of a control's operation are classified as compliance tests. Procedures that involve determining whether the transaction was correctly processed with respect to the monetary amount are classified as substantive tests. Many of the procedures in this latter group are, in fact, dual-purpose tests. The compliance aspect of these procedures is limited to instances where monetary amounts that are incorrectly processed provide evidence that a particular set of control procedures are not effectively operating.

The tentative list of principal substantive tests of the accounts receivable balance —including tests of details and overall tests—are as follows:

Code	Description
P1	Request confirmation of the recorded amounts
P2	Analyze and test the account from the date of confirmation to the closing date
P3	Review classification and disclosure
P4	Test aging
P5	Test subsequent collections
P6	Test period cutoffs of sales and inventory
P7	Analyze and evaluate allowance for doubtful accounts
P8	Reconcile accounts receivable detail file to general ledger balances
P9	Reconcile total credit to sales to debit to accounts receivable
P10	Reconcile sales order volume to capabilities, capacities, etc.

The following matrix demonstrates the interdependence between the set of substantive tests of balances and the previously cited audit objectives.

	Procedures									
Objectives	P1	P2	P3	P4	P5	P6	P7	P8	P9	P10
1	x	x		x	x	x	x	x	x	x
2						x	x		x	x
3			x							
4	x	x				x			x	x
5						x			x	x

Note that procedures that test for overstatements of sales (4) are included among the procedures that test for an overstatement of accounts receivable (1). This reflects the fact that the primary test for an overstatement of sales is testing for an overstatement of accounts receivable. Likewise, tests for an understatement of accounts receivable are included among those that directly test for sales understatement (except for analyzing and evaluating the allowance for doubtful accounts). Any evidence of understatement of accounts receivable discovered through the confirmation procedure is useful, but the confirmation procedure cannot be regarded as a major source for discovering understatement errors.

Identifying Pertinent Accounting Controls

The pertinent accounting controls are identified with the aid of the interdependence matrix. First the auditor identifies the major sources of errors and irregularities. These are shown in the first column on the left. Next, using the internal control questionnaire together with the flowcharts, the key controls are identified and listed as shown in the second column of the interdependence matrix.

Determining the Maximum Degree of Reliance

The third column indicates the maximum possible reliance on these key controls. The auditor uses a qualitative scale consisting of four categories: high, moderate, low, and none. For each set of pertinent controls, the auditor determines the maximum possible reliance, assuming that compliance with the controls is satisfactory.

Determining the Extent and Timing of Substantive Tests

The auditor uses the interdependence matrix as an aid for determining the extent of substantive procedures. The next to last column shows, for each type of error or irregularity, the planned substantive and dual-purpose procedures. Referring to the third column on the left, the auditor could easily determine the maximum possible degree of reliance corresponding to each of the planned substantive procedures. The auditor decides to consider the extent of substantive tests at several possible degrees of reliance. To illustrate this process for the Maxwell case, the major objectives concerning accounts receivable overstatement and sales understatement are considered in detail. In both cases, the allowable sampling portion of the overall audit risk ($1 - RS$) is taken as .05.

Accounts Receivable Overstatement. The planned substantive tests of details of the accounts receivable balance include requesting confirmation of a sample of recorded amounts, testing the aging, testing subsequent collections, and testing period cutoffs. Of these, only the confirmation requests are to be done statistically.

The interdependence matrix shows that the controls pertinent to confirmation requests are regarded as highly effective in preventing material errors from occurring. Consequently, the auditor's maximum possible reliance on the pertinent controls is quite high and he considers the extent of his sample for degrees of reliance corresponding to none, low, moderate, and high.

Materiality—The auditor has previously determined that $50,000 would represent a material amount for accounts receivable. The tests of details need to be planned so that, when considered together, they provide evidence concerning the possible existence of this amount of overstatement. To accomplish this objective, the auditor needs to specify how the material amount ($50,000) is to be divided between the nonstatistical tests and confirmation requests. In this case, Maxwell's auditor judgmentally determines that $20,000 would represent his outside limit on the amount of monetary error that could remain undetected in his planned tests of aging, subsequent collections, and period cutoffs. The remaining $30,000 ($50,000 − $20,000) could then be used as the measure of auditing materiality in designing the sample for requesting confirmations.

Sampling risk—The auditor considers several possible levels for his tolerable sampling risk corresponding to the possible degrees of reliance. Employing the suggestions made in chapter 7, the auditor uses the following correspondence between the tolerable sampling risk and the degree of reliance on the pertinent internal accounting controls.

Degree of Reliance	Tolerable Sampling Risk
None	.07
Low	.10
Moderate	.15
High	.25

These values reflect the auditor's planned use of analytical review procedures with a reliance subjectively evaluated at .33 and maintain the sampling portion of the audit risk at .05.

Risk of overauditing—To determine the tolerable risk of overauditing, the auditor considers the sources of evidence that might be used in the event the sample indicates there could be a material misstatement of accounts receivable. The auditor concludes that requesting additional customers' confirmations represents the best source and, on that basis, decides the risk of overauditing should be kept low. Accordingly, .05 is selected as the tolerable level for this risk.

Sample design—The computer record of the accounts receivable is a convenient frame, and the customer account balance is designated as the sampling unit. The computer program (CPA-1) is run to create a frequency distribution of the recorded amounts. This shows that only 20 customers had balances exceeding $800. There are no customers with zero or credit balances. There are 4,500 customers who have balances amounting to less than $800. The total recorded amount is $1,400,000, and the total of accounts above $800 amounts to $150,000.

To decide between using a pps sample or a stratified random sample, the auditor considers the possibility that he will observe any differences between recorded and audited amounts. Based on his experience during the previous year, he believes this unlikely and consequently decides to use a pps sample.

The sample sizes are determined from Kaplan's tables using a material fraction of .0214 (30,000/1,400,000) corresponding to a material amount of $30,000 and a total recorded amount of $1,400,000. For each considered sampling risk, the risk of over-auditing is set at .05 at a fraction of .005 of the account balance or $7,000. The resulting sample sizes following the procedure described in Appendix 6A are as follows:

Tolerable sampling risk	.07	.10	.15	.25
Sample size	478	374	342	239

The decision concerning which of these to use is made after the required compliance tests are tentatively planned.

Sales Understatement. Because hash totals of customer order numbers are prepared after processing, there is a possibility that documents could be lost prior to keypunching. As a consequence of this control weakness, the auditor regards the test of bills of lading primarily as a substantive test. The auditor also plans to review the on-order file for evidence that orders could be shipped without being recorded as a sale. These two tests, together with the test of period cutoffs of sales, constitute the substantive tests of details pertaining directly to sales understatement. Tests made in other parts of the audit (for example, inventory) also provide information concerning understated sales, but such sources of evidence are not considered in this case study.

Materiality—The tests of detail taken together should provide evidence concerning whether there might be a material amount of understatement. The auditor has previously decided that $50,000 would represent a material amount.[2] How should that amount be allocated among the tests of details?

Concerning the volume of daily sales (about 200), the auditor decides that $10,000 would be an appropriate outside limit on the amount of undetected error associated with his planned test of period cutoffs. This leaves $40,000 for the test of the bills of lading and the review of the on-order file. No allocation of the $40,000 between these two tests is required because each examines the same source of error.

Sampling risk—Since the auditor's planned reliance on the system of controls designed to prevent or detect unrecorded sales is very low, he sets the tolerable sampling risk at .10 for the tests considered together. This reflects his determination that the analytic review of monthly and year-to-year sales is somewhat effective (about .33) for detecting any material understatement.

2. It is possible to use a different material amount for understatement and overstatement. In this case study, the amounts were equal.

The test of the bills of lading and the review of the on-order file both have the same objective, and thus the risk that together they fail to produce evidence of an existent material error is the product of their individual risks. Since the review of the on-order file is not a statistical test, its risk cannot be objectively measured, but the auditor decides that he can subjectively value the risk.

Before making a final decision on the appropriate scope for each of these tests, the auditor tentatively plans to use a .50 sampling risk for the bills-of-lading test together with a .20 risk for the review of the on-order file ($.50 \times .20 = .10$).

Sample design—To determine the required sample size for the test of the bills of lading at a .50 sampling risk, the auditor considers the likely results based upon his previous experience with Maxwell. He anticipates few, if any, differences and consequently decides the sample should be sufficiently large to provide a useful upper limit for the total amount of difference between shipments and recorded sales.

Since the sample will be selected from the bills-of-lading file, there is no opportunity to stratify the population nor to use pps selection. Consequently, an unrestricted random sample of the bills of lading numbers is the only feasible method.

The auditor uses $500 to represent the largest shipment amount. He justifies this by reviewing the sales activity for the few customers (about 30) who accounted for nearly all the orders over $500. To determine the sample size, he calculates the percent of the total shipments (50,000) required to equal a material amount ($40,000) when each shipment is at the maximum value ($500).[3] Solving the equation

$$(50{,}000p)\ \$500 = \$40{,}000,$$

he finds $p = .0016$ or .16 percent. To obtain an upper precision limit of .0016 for the proportion of sales when no differences are observed, he uses Kaplan's tables to find that 432 observations are required to achieve a tolerable sampling risk of .50 (reliability .50). This large sample size is the result of using a very inefficient sample design, but, unfortunately, a more efficient one is not feasible in these circumstances.

Note that the risk of overauditing is not explicitly considered in determining the required sample size. The reason for this is simply that the desired upper limit of .0016 percent is too low to permit considering the risk of overauditing.

The auditor decides that such a large sample for the bills-of-lading test is not feasible and consequently determines that the scope of his other procedures will have to be sufficient to provide the required evidence pertaining to potential sales understatement. He concludes that a thorough review of the on-order file, together with a careful examination of the customer order form file for gaps in the number sequence, will provide sufficient evidence. In addition, he plans to broaden the scope of his observation of inventory quantities.

Determining the Extent and Timing of Compliance Tests

The statistical compliance tests required to support the considered degrees of reliance are grouped within three separate populations: customer orders, credit memos, and suspense report items. The interdependence matrix shows that the test of compliance with respect to credit memos (T12) may be considered separately from the other two. On the other hand, the test of customer orders and the test of suspense

---

3. Using the maximum amount ($500) produces the most conservative upper limit. Using another amount, such as the average shipment amount, might in some situations be appropriate, depending upon the distribution of sales and other evidence.

report items are planned jointly to corroborate the auditor's planned reliance on those pertinent controls that affect the overstatement of accounts receivable.

The fact that statistical sampling is used does not affect the timing decisions. In the Maxwell case, the auditor decides to conduct his tests of transactions—both compliance and substantive—during the first weeks of June.

Credit Memos. Considering the test of credit memos, the auditor decides that his planned reliance will be relatively low. In addition, he determines that returns and allowances amounted only to about $300,000 or 3 percent of total sales. Consequently, nearly one-sixth of the dollar amount of these transactions will have to be incorrect to produce a material error. Recognizing the variability among sale amounts the auditor decides that .04 will represent a conservative value for the threshold rate.

To determine a sample size, the auditor notes from the attribute tables that a sample of 40 would provide an upper precision limit of .14 at .90 reliability even when the anticipated occurrence rate is as high as .05. Since his firm's policy on statistical sampling requires special approval for using a sample of less than 50, he prepares a memo setting forth the circumstances and promptly receives permission to use 40. Because this test is not closely related to the other statistical tests, it is not developed any further in this chapter.

Customer Orders and Suspense Report Items. Planning the statistical compliance tests involving customer orders and suspense report items is done jointly to reflect the possibility that a material amount of error could occur because either sales or cash receipts are improperly processed. For the Maxwell case, the auditor decides that because the suspense account serves only as a memo record for unallocated cash receipts, there is no likelihood that a material misstatement could be produced. Following this decision, his planning concentrates on the tests of customer orders.

Risk of unwarranted reliance—Corresponding to each of the considered degrees of reliance, the auditor specifies the tolerable risk of unwarranted reliance. For the confirmation procedure, the risks are as follows.

Reliance	Tolerable Sampling Risk	Tolerable Risk of Unwarranted Reliance
None	.07	—
Low	.10	.33
Moderate	.15	.12
High	.25	.05

These were determined as described in chapter 7 so that any increase in the sampling risk above the risk corresponding to no reliance (.07) multiplied by the risk of unwarranted reliance should not exceed .01 [(.10 − .07) × .33 = .01, (.15 − .07) × .12 = .01, (.25 − .07) × .05 = .01].

As a result, the compliance tests for those pertinent procedures relate to the confirmation requests planned at .05, .12, and .33 corresponding to high, moderate, and low degrees of reliance.

In addition, the auditor considers his levels of reliance with respect to the nonstatistical tests of details. He regards his planned degree of reliance as low for each of the nonstatistical tests except for the test of aging and the review of subsequent collections. For these tests, he regards his reliance on sales' being properly approved or authorized as moderate. Consequently, he decides to use .10 as the risk of unwarranted reliance relative to this pertinent procedure.

Risk of overauditing—Because the costs of increasing (1) the number of confirmation requests to the level required for no reliance (478) and (2) the extent of his aging test and because review of subsequent collections is not high, the auditor does not explicitly consider the risk of overauditing in his planning.

Threshold rates—For the customer orders, the pertinent controls to be statistically tested concern whether the order was properly approved or authorized, whether prices and quantities were checked for accuracy, and, if paid, whether the clerks check that the amount paid agrees with the amount shown on the remittance advice card and whether the cashier agrees the checks to the list of remittances. For each of these pertinent controls, the auditor needs to determine the threshold rate for nonsatisfactory compliance.

Proper approval or authorization primarily affects the collectibility of the account. There are approximately 50,000 sales transactions within the year. The actual number may vary from this, but since the 6,000 customers order an average of eight times per year, the total should be reasonably close to 50,000. A very conservative candidate for a threshold rate is .5 percent. This is based on the observation that if fewer than 250 transactions (.005 × 50,000) at the average amount of $200 were not authorized or approved, less than a material amount of sales ($50,000) would be involved. Considering the nature of the controls, the number of new customers involved and the likelihood that all sales to a customer within a three-month period would be made without proper authorization, the auditor judges that the threshold rate can safely be set higher than .5 percent. Taking all factors into account, he determines that 1.5 percent would be an appropriate threshold rate.

The threshold rate for checking the accuracy of prices and quantities is determined to be 2 percent. This rate is based on the auditor's judgment that errors in prices and quantities will not likely exceed 25 percent of the order. Consequently, a material number of transactions would be 1,000 because it would require 1,000 transactions at the average sale of $200 containing an error of $50 (.25 × $200) to reach a material amount of monetary error ($50,000).

To determine the threshold rate for the pertinent controls over cash receipts to be tested statistically, the auditor observes that 6,000 customers ordering an average of eight times per year would generate about 50,000 remittances during the year. Considering the average payment of $200 and regarding a discrepancy of more than 25 percent between check amount and recorded amount as highly unlikely, the auditor determines that the threshold rate will be 2 percent ($50,000 = 48,000 × $210 × .25 × .02).

Sample sizes—Having determined the threshold rates for each of the three pertinent procedures to be tested statistically the auditor then uses the attribute tables to look up the sample sizes required for each considered degree of reliance.

For proper approval or authorization, the threshold rate is .015, and the tolerable risk of unwarranted reliance is .10. Because the auditor anticipates no compliance deviations, the sample size is determined to be 153 for this attribute.

For checking the accuracy of prices and quantities, the threshold rate is .02. From the previous year's experience, the auditor anticipates about a .004 rate of compliance deviation. The required sample sizes are as follows:

Sample size	235	182	115
Risk of unwarranted reliance	.05	.12	.33

For checking whether the amount paid agrees with the amount owed, the threshold rate is also .02. The auditor regards the two procedures performed by the cash receipts clerks and the cashier as a single procedure designed to insure the correctness of the recorded receipts. Since the auditor has not previously found evidence of any compliance deviation from either procedure, he uses an anticipated rate of occurrence equal to zero. The following represents the required sample sizes:

Sample size	150	105	55
Risk of unwarranted reliance	.05	.12	.33

Determining the Planned Degree of Reliance

At this point in the planning process, the auditor determines the sample sizes for his compliance tests and confirmation requests corresponding to three potential degrees of reliance—low, moderate, and high. These sample sizes are as follows:

	Degree of Reliance			
	None	Low	Moderate	High
Confirmation requests	478	374	342	239
Customer orders	153	153	182	235

The sample sizes for the customer orders correspond to the largest sample required for testing compliance with any of the pertinent controls. The 153 observations shown under the condition of no reliance refer to the sample size required to test whether orders receive proper approval or authorization. This reflects his decision that his planned audit program to test aging and collectibility requires moderate reliance on this control. In practice, the auditor might consider the possibility of reducing this reliance and increasing the scope of his tests of aging and collectibility. This has not been done here.

To determine his planned degree of reliance, the auditor considers the costs of the four alternatives. He uses $2 to represent the cost of auditing a confirmation request that was returned and $5 to represent the cost if the request was not returned. From his experience, he anticipates that about 30 percent of the requests will not be returned. Using these numbers, he calculates the expected cost corresponding to each potential degree of reliance. For example, the expected cost of sending 478 confirmation requests is $1,386 (478 × .30 × $5 + 478 × .70 × $2).

The costs are as follows.

	Degree of Reliance			
	None	Low	Moderate	High
Confirmation requests	478	374	342	239
Expected cost	$1386	$1085	$992	$693

Noting that increasing the degree of reliance from low to moderate entails 29 additional sample items and from moderate to high required 53 more items, the auditor calculates the increased cost for each step. He determines that 29 additional items would add from $50 to $70 to the cost and 53 additional items would add another $100 to $125. Consequently, he decides that using the high degree of reliance would be the most economical combination.

Final Design of Compliance and Dual-Purpose Tests

Following the preliminary design of the audit program, the auditor needs to design the compliance and dual-purpose tests so that they can be accomplished in the most efficient manner. In the Maxwell case, the auditor decides that since the order numbers are noted on the filed bills-of-lading copies in the shipping department, bills-of-lading files can be used as a frame to test compliance with pertinent procedures concerning sales and receipts. Specifically, for the selected sample of bills-of-lading numbers, the auditor would obtain an order number and trace this to a daily sales report (if possible). For the required number of those customer orders on the daily sales reports, the auditor would perform the indicated tests of compliance. Any order number not appearing on a daily sales report would be evidence of goods shipped without being recorded as a sale and, of course, would provide additional evidence concerning sales understatement.

The sampling plan is as follows.

Population: All shipments from the beginning of the year to May 31.

Frame: Shipping department files of bills of lading.

Objectives:

1. To test whether compliance is satisfactory with respect to the following attributes:
   a. Customer order form is initialed by an order clerk, Mr. Jones or Mr. Gordon, to indicate the order was authorized or approved.
   b. Customer order form is initialed by accounting clerk to indicate that quantities on order form, bill of lading, and sales report match and that price corresponds to current price list.
   c. Cash receipts clerk marks listing to indicate check amount agrees with remittance advice card and cashier initials listing to indicate check amount agrees with listing.
2. To ascertain, on a limited basis, whether shipments are made without being billed as sales.

Sample size: The required sample sizes for each objective are as follows.

Objective	Sample Size	Planned Upper Precision Limit	Reliability
1a	153	.015	.90
1b	235	.02	.95
1c	150	.02	.95
2	N/A	N/A	N/A

To accomplish this, 300 random numbers are selected from the bills-of-lading numbers used during the period. These are recorded in order of selection and the first 235 with an order number are used for 1b, the first 153 for 1a, and the first 150 of those paid for 1c. The auditor selects 300 to provide for the possibility that some bills of lading may not have a corresponding order form and to provide at least 150 paid orders. This is proper procedure from a statistical viewpoint, but the auditor may decide to examine all 235 for compliance with the pertinent procedures. In that event, the auditor would be doing more than the required work corresponding to high reliance.

Audit procedures:

1. Trace each selected bill of lading to a sales order in the accounting department completed order file. Note quantity shipped, quantity ordered, and difference. If no order number shown, difference is quantity shipped.
2. For those selected for additional testing, observe whether customer order form is initialed by an order clerk, Mr. Jones or Mr. Gordon, to indicate the order was authorized or approved.
3. For those selected for additional testing, observe whether accounting clerk initialed order to indicate that quantities were matched and prices checked. Make an independent comparison of quantities and prices.
4. For selected orders, trace to monthly customer statement accounts receivable trial balance, and if paid or removed via a credit, apply procedure 5. If not paid, trace to the open order files.
5. Trace payment to cash accepted-cash rejected report and receipted duplicate deposit slip. Trace to listing of remittance advice cards and observe whether cash receipts clerk marked listing to indicate agreement between check amount and remittance advice card and whether cashier initialed listing to indicate check amount agrees with listing. Examine following aged trial balance to see whether receipt has been credited to customer's account. If removed from account by other means (for example, credit memos), examine support and recording.

Some of these audit procedures are not part of the statistical compliance test (such as testing whether a receipt has been credited to the customer's account), but represent procedures that are part of the substantive tests of transactions. Whether these substantive tests of transactions are done statistically or judgmentally, they should be planned so that the sample is large enough to enable the auditor to reach a useful conclusion. In the Maxwell case, the auditor determines that examining 150 orders that were paid provides sufficient evidence regarding the payment being credited to the right customer.

Execution and Evaluation of Compliance and Dual-Purpose Tests

The planned statistical tests of transactions are done during the early part of June. The results of this and other nonstatistical compliance tests are then evaluated to determine whether the planned degree of reliance is warranted or whether the tentative audit program needs to be changed.

The planned statistical tests have both substantive and compliance objectives (dual-purpose) and need to be evaluated accordingly. For example, if *any* of the 235 randomly selected bills of lading fails to indicate a recorded sale, the auditor would require additional evidence to satisfy himself regarding the possibility that sales were understated. The work papers should clearly indicate the additional audit procedures to be employed should that be necessary.

The compliance aspect is planned so that if there are any instances of lack of approval or authorization, the degree of planned reliance will be reduced and the scope of the audit concerning collectibility increased. Any required modification to the preliminary audit program should be clearly indicated in the work papers.

The statistical test of compliance with respect to the sales orders and cash receipts directly affects the planned timing and extent of the requests for accounts receivable confirmation. If there is more than one occurrence of no initial by the accounting clerk or any occurrence of the cash receipts clerk or cashier failing to mark or initial the listing, the auditor's planned reliance on the pertinent controls will be reduced. In this case the auditor decides that if more than the stipulated number of occurrences is observed, he will increase the number of confirmation requests from the planned 223 to 478. In addition, a qualitative evaluation of the cause of any observed deviations will be made, and if there is any indication of potential difficulty, the timing of the requests will be moved closer to the end of the year.

The planned audit procedures that are not part of the statistical compliance test are to be judgmentally evaluated.

In the Maxwell case, the planned compliance tests, both statistical and non-statistical, do not reveal any evidence of compliance deviations exceeding the allowable limits, and, consequently, the preliminary audit program is adopted as the audit program. The substantive tests of transactions likewise reveal no evidence of potential material monetary errors.

Final Design of Substantive Tests

Since the compliance test corroborates the auditor's planned reliance, the auditor designs the statistical confirmation requests with a tolerable sampling risk of .25, a risk of overauditing of .05, and a material amount of $30,000. As indicated in the preliminary design, a pps sample of 239 confirmation requests is appropriate in these circumstances. Had the auditor discovered evidence through his compliance tests of potential differences in the recorded amounts of accounts receivable, not only would the sample size be increased, but a different statistical method might be used. For Maxwell, the details of the final design are as follows:

Population: All recorded accounts receivable having a positive balance as of June 30. Total recorded amount equals $1,400,000.

Frame: Computerized accounts receivable master file.

Sampling unit: A customer account balance.

Sampling risk: .25

Risk of overauditing: .05 (at a fraction of .005 of account balance).

Materiality: $30,000. Material fraction: .0214 (30,000/1,400,000).

Sample size: 239

Selection method: pps

This sample size permits observing three errors without exceeding the material amount.

Execution and Evaluation of Confirmation Requests

The 239 confirmation requests are selected and mailed under the auditor's supervision. Within ten days, 104 are returned and at that time second requests are mailed to the remaining 135 customers. The second request results in 47 additional responses. Telephone requests to the remaining 88 result in twelve additional responses. For the remaining 76 requests, the auditor performs alternative procedures.

Had the audited amount agreed with the recorded amount in at least 236 of the 239 sampling units, the auditor would have concluded that his substantive test of details indicated no material amount of error in the accounts receivable balance with respect to existence or proper recording. Any exception would, of course, be investigated to determine the cause.

In Maxwell's case, the results are that nine of the 239 requests indicate a difference. In each case, investigation shows that a new product line had been introduced in April and as an introductory offer, established customers could order up to $300 of the new line on consignment. To evaluate the potential effect of this on the financial statements, the auditor first determines that, on the basis of observing nine in 239, the upper precision limit at .95 reliability would be about .07. Consequently, from the sample evidence, he concludes that as many as 420 such consignments could have been made (420 = 6000 × .07). Since each consignment could have been for $300, the potential overstatement is $126,000 ($300 × 420). This amount would represent material overstatement, and, hence, the auditor needs to obtain additional information.

As a result, the auditor decides to investigate the shipments for the new product line by examining the activity of those products in the inventory records. Establishing the amount and prices of the units shipped would allow the auditor to obtain a reasonable estimate of the extent of overstatement of accounts receivable as well as the understatement of the inventory. This is done and the results indicate that the accounts receivable were overstated by $60,000 while inventory was understated by $40,000. An adjustment to the records is then made by the client.

Summary

This case study illustrates how statistical sampling can be integrated into the auditing process. The integration begins with the preliminary design of the audit program. In this phase of the audit, the auditor decides the material amount ($50,000) to use for the accounts receivable/sales portion of the audit. He uses this together with his overall tolerable sampling risk (.05) to determine, on a preliminary basis, the extent (sample size) of each of his planned statistical tests—both substantive and compliance—at several possible levels of reliance. He then compares the costs of the possible alternatives and chooses the most economical one. In the Maxwell case, a potential substantive test of the bills of lading is found to require too many observations to be useful and is replaced by other sources of evidence.

The final design and execution of the compliance and dual-purpose tests of transactions are described. Because the evidence corroborates the auditor's preliminary evaluation of the system, no changes are required for the statistical test of the accounts receivable balance.

The evaluation of the confirmation requests reveals a situation that has not been anticipated during the planning phase. This is typical of many auditing situations. Using both statistical and qualitative evaluation, the auditor is able to resolve the potential problem.

Had the auditor anticipated observing differences from the recorded amounts in accounts receivable he might have used a different statistical design. One possibility would be to stratify the population by size of the recorded amounts and compute a sample size required to achieve the tolerable sampling risk at the specified material amount and the desired risk of overauditing. In the Maxwell case, the tolerable sampling risks considered are .07, .10, .15, and .25, with the material amount equal to $30,000. The risk of overauditing is equal to .05. Using the formula for determining the desired standard error, assuming that the negative approach is employed, the following represents the desired standard error at the four risk levels.[4]

Desired standard error	$8,720	$9,259	$10,000	$11,364
Tolerable risk	.07	.10	.15	.25

These amounts would then have been used in the computer programs (CPA-1 and Plan 2, described in chapter 9) to determine the required sample size corresponding to each risk level.

4. The formula is

$$\text{standard error} = \frac{M}{U_{R_1} + z_{\alpha/2}},$$

where U_{R_1} represents the desired one-sided reliability factor, and $z_{\alpha/2}$ represents the normal factor corresponding to the risk of overauditing. In this example, $z_{\alpha/2} = 1.96$. $U_{R_1} = 1.48$ corresponding to a risk of .07, $U_{R_1} = 1.28$ corresponding to .10, $U_{R_1} = 1.04$ corresponding to .15, and $U_{R_1} = .68$ corresponding to .25.

9

Computer Programs

The computer has proved to be an invaluable aid to auditors who use statistical sampling techniques. Computer programs have been written that enable the auditor to apply the statistical methods described in the earlier chapters of the book. These programs are of two types: programs that perform statistical routines on files of machine-readable data and time sharing programs. There are three programs of the first type (called batch mode) that are designed to be run together with an edited version of the client's file of recorded amounts. One of the programs provides a profile of the recorded population, and the other two select either a stratified random sample or a pps sample from the recorded population. These batch programs are written in ANSI COBOL.

All the time sharing programs are written in the BASIC language. The programs are of two general types: planning and evaluation. There are four programs that can be used in planning. Two of these compute the required sample size for an attribute sampling plan—either estimation, decision, or discovery. The other two are to be used for variable sampling. One of these computes the planned precision and reliability corresponding to specified risks and material amount, and the second determines the location of stratum boundaries, the required sample size, and the allocation of the sample to the strata.

Eight time sharing programs assist the auditor in evaluating the results of the statistical sample. One of these pertains to attributes. It calculates either an achieved precision limit or an achieved precision interval. The other seven pertain to variables. Six of these calculate the estimated audited amount together with the achieved precision using one of seven basic methods—stratified mean estimation, stratified difference estimation, combined ratio estimation, combined regression estimation, separate ratio estimation, separate regression estimation, and pps estimation. The other statistically evaluates the estimated total audited amount together with the achieved precision from the decision standpoint—does the statistical evidence support the proposition that no material error exists?

Copies of the three batch and twelve time sharing programs are available for a nominal charge from the Computer Services Division of the AICPA.

Each of the programs is briefly described in terms of its purpose, input, process, and output. Following the description, there is a worked example of each of the programs. In each of the time sharing evaluation programs, the degrees of freedom are calculated using a formula in Cochran [5], p. 95, based on the number of observed differences. The resulting number of degrees of freedom is never larger than the number of differences minus the number of strata. This represents a more conservative result than described in chapter 6.

ATSIZ1

Purpose: To calculate the sample size required to achieve a specified one-sided reliability that the occurrence rate of an attribute is less than a specified threshold rate.

Input: In response to a message printed at the computer terminal, the user specifies the population size (N), threshold rate (P_0), one-sided reliability (R_1), and anticipated occurrence rate in the sample ($AØ$). P_0, R_1, and $AØ$ are expressed in decimals.

Process: The program uses the hypergeometric distribution to find the smallest sample size required to make the probability equal to or greater than the one-sided reliability (R_1) of observing a sample occurrence rate larger than the anticipated sample occurrence rate ($AØ$) when the population occurrence rate is P_0.

Output: The output includes the required sample size and the critical number of occurrences. Should the number of occurrences exceed the critical number of occurrences, the achieved upper precision limit will exceed the threshold rate.

In addition, for each number of occurrences up through the critical number, the program prints the probability of observing no more than that number when the population rate equals the threshold rate (P_0). These probabilities are expressed in percentages.

Example: Two examples of the operation of the ATSIZ1 program are presented on the following page.

The first example illustrates the use of the program to obtain an upper precision limit of .05 at .95 one-sided reliability if the occurrence rate in a population of 100,000 is .01. The example was discussed in the attribute estimation section of chapter 4. As indicated by the printout, the required sample size is 93 and the critical number of occurrences is one. If one occurrence is found, the sample will provide a 95 percent reliability (100 − 5 percent) that the occurrence rate in the population is less than 5 percent. If no occurrences are found, the reliability will be more. If more than one occurrence is found, the auditor will have less than the desired 95 percent reliability that the actual occurrence rate does not exceed the 5 percent threshold rate.

The second example illustrates the use of the program for discovery sampling. The example is discussed in the discovery sampling section of chapter 4. The printout indicates that a sample of 298 will be required to provide 95 percent probability of observing at least one occurrence when the population occurrence rate is .01.

```
>RUN
14:19   DEC  9  /ATSIZ1

INPUT  IS IN DECIMALS, OUTPUT IS IN PERCENTAGES.

TYPE THE POPULATION SIZE (N), THRESHOLD RATE (P0), ONE
SIDED RELIABILITY (R1) AND ANTICIPATED OCCURRENCE RATE IN
THE SAMPLE (A0)
N,P0,R1,A0=   ?100000,.05,.95,.01

REQUIRED SAMPLE SIZE IS    93
CRITICAL NUMBER OF OCCURRENCES IS    1

THE EVALUATION OF THE RESULTS WILL DEPEND ON THE ACTUAL
NUMBER OF OCCURRENCES IN THE SAMPLE, AS FOLLOWS:

NUMBER OF      PERCENT OF     PERCENT PROBABILITY OF A
OCCURRENCES    OCCURRENCES    RATE THIS SMALL IF  POPULATION
IN SAMPLE      IN SAMPLE      OCCURRENCE RATE WERE   5  %
 0              0               .845896
 1              1.07527         4.99035


     630 HALT
>


>RUN



14:20   DEC  9  /ATSIZ1

INPUT  IS IN DECIMALS, OUTPUT IS IN PERCENTAGES.

TYPE THE POPULATION SIZE (N), THRESHOLD RATE (P0), ONE
SIDED RELIABILITY (R1) AND ANTICIPATED OCCURRENCE RATE IN
THE SAMPLE (A0)
N,P0,R1,A0=   ?100000,.01,.95,0

REQUIRED SAMPLE SIZE IS    298


     630 HALT
>
```

ATSIZ2

Purpose: To calculate the sample size required to achieve a tolerable risk of deciding that the population occurrence rate is less than a specified threshold rate (P_0) when it is P_0 and a tolerable risk of deciding that the population occurrence rate is more than P_0 when it is P_1, a specified lesser acceptable occurrence rate.

Input: In response to messages printed at the computer terminal, the user specifies the population size (N), the threshold rate (P_0), a lesser acceptable occurrence rate (P_1), the tolerable risk of deciding that the population occurrence rate is less than P_0 when it is P_0 (the tolerable risk of unwarranted reliance), and the tolerable risk of deciding that the population occurrence rate exceeds P_0 when it is P_1 (the tolerable risk of overauditing). These risks together with P_0 and P_1 are expressed in decimals.

Process: The program uses the hypergeometric distribution to find the smallest sample size and critical number of occurrences which satisfy the following two conditions:

1. The probability of the number of sample occurrences being less than or equal to the *critical number* is no greater than the tolerable risk of unwarranted reliance when the occurrence rate is P_0.
2. The probability of the number of sample occurrences exceeding the critical number is no greater than the risk of overauditing when the occurrence rate is P_1.

Output: The output includes the required sample size and the critical number of observed occurrences.

In addition, for each number of occurrences up through the critical number, the program prints the probability of observing no more than that number when the population rate equals the threshold rate (P_0). These probabilities are expressed in percentages.

Example: An example of the operation of ATSIZ2 is shown on the following page. The example was discussed in the attribute decision section of chapter 4. If the population size is 100,000 and the auditor specifies a .05 risk of unwarranted reliance at $P_0 = .05$ and a .05 risk of overauditing at $P_1 = .01$, the required sample size is 181 and the critical number of occurrences is four. There is a 4.9 percent probability of observing no more than four occurrences when the population occurrence rate equals the 5 percent threshold rate. On the other hand, if the population occurrence rate is .01, the probability of finding more than four occurrences is less than .05.

```
>


>RUN
14:16   DEC  9  /ATSIZ2

INPUT  IS IN DECIMALS, OUTPUT IS IN PERCENTAGES.

TYPE THE POPULATION SIZE (N), THRESHOLD RATE (P0) AND
A LESSER ACCEPTABLE OCCURRENCE RATE (P1)
N,P0,P1=  ?100000,.05,.01
RISK OF DECIDING THAT POPULATION OCCURRENCE RATE IS LESS
THAN P0 WHEN IT IS P0=  ?.05
RISK OF DECIDING THAT POPULATION OCCURRENCE RATE IS MORE
THAN P0 WHEN IT IS P1=  ?.05

REQUIRED SAMPLE SIZE IS    181
CRITICAL NUMBER OF OBSERVED OCCURRENCES IS   4

THE EVALUATION OF THE RESULTS WILL DEPEND ON THE ACTUAL
NUMBER OF OCCURRENCES IN THE SAMPLE, AS FOLLOWS:

NUMBER OF     PERCENT OF    PERCENT PROBABILITY OF A
OCCURRENCES   OCCURRENCES   RATE THIS SMALL IF  POPULATION
IN SAMPLE     IN SAMPLE     OCCURRENCE RATE WERE   5  %
 0             0             9.20968E-03
 1             .552486       9.71105E-02
 2             1.10497       .514186
 3             1.65746       1.82588
 4             2.20994       4.90190


     780 HALT
>
```

ATEVAL

Purpose: To calculate either the achieved upper precision limit at a specified one-sided reliability or the achieved precision interval at a specified interval reliability (two-sided).

Input: In response to messages printed at the computer terminal, the user specifies population size, sample size, and number of sample occurrences. The user also specifies the reliability level and designates whether it is one-sided or two-sided (interval).

Process: The program uses the hypergeometric distribution to find the appropriate precision limits. For the one-sided case, the program searches for a population occurrence rate that makes the probability equal to the one-sided reliability of observing more than the observed number of occurrences. For the interval (two-sided) case, the program searches for two population occurrence rates. The upper limit corresponds to that rate which makes the probability equal to $(1 + R)/2$ of observing more than the observed number of occurrences. The lower limit corresponds to that rate which makes the probability equal to $(1 + R)/2$ of observing fewer than the observed number of occurrences.

Output: The program prints the sample occurrence rate together with the achieved upper precision limit for the one-sided reliability or both the achieved lower precision limit and upper precision limit for the two-sided (interval) reliability.

Example: Two examples of the operation of the ATEVAL program are shown on the following page. The first example illustrates the calculation of achieved precision at a one-sided reliability, using the sample results described in the attribute estimation section of chapter 4. The second example illustrates the calculation of two-sided reliability using the results described in the same section.

```
>RUN

14:13   DEC  9  /ATEVAL

POPULATION SIZE, SAMPLE SIZE, NUMBER OF OCCURRENCES IN SAMPLE=
?100000,120,1
PERCENTAGE CONFIDENCE LEVEL
?95
1 OR 2 SIDED  ?1

OCCURRENCE RATE IN SAMPLE=    .833333  %
  UPPER PRECISION LIMIT=   3.90000  %


    1180 HALT
>




>RUN


14:14   DEC  9  /ATEVAL

POPULATION SIZE, SAMPLE SIZE, NUMBER OF OCCURRENCES IN SAMPLE=
?100000,120,4
PERCENTAGE CONFIDENCE LEVEL
?95
1 OR 2 SIDED  ?2

OCCURRENCE RATE IN SAMPLE=   3.33333  %
  LOWER PRECISION LIMIT=   .912500  %
  UPPER PRECISION LIMIT=   8.35000  %


    1180 HALT
>
```

CPA1

Purpose: To obtain a profile of the population of recorded amounts by means of a frequency distribution.

Input: Input consists of a record file and one or more specification cards.

The record file is the file from which the sample will be selected. The file may be on punched cards, magnetic tape, or disk storage. Each record in the file must be in the following format:

Position(s)	Description
1–9	Transaction amounts in cents, right justified. In other words, 000015000 is interpreted as $150.00
10	Blank for a positive transaction, or a minus sign (−) for a negative transaction.
11–80	Description of transaction. Any combination of alphabetic and numeric characters is acceptable.

Because the file must be in this fixed format, it will be necessary to prepare a program to write the record file from the client's actual data file.

Specification cards are used to divide the total range of recorded transaction amounts into specified ranges and, within each range, into equally spaced cells. Separate cards must be used for positive and negative ranges. The total number of cells for positive amounts may not exceed 100 and the total number for negative amounts may not exceed 100. The format for each specification card is as follows:

Card Column(s)	Description
1	*P* to indicate that the transaction amounts on the card are positive or *N* to indicate they are negative.
2–10 13–21 24–32 35–43 46–54 57–65 68–76	Maximum transaction amount in each range, beginning with the lowest amount in columns 2–10 and proceeding to successively higher amounts in columns 13–21, 24–32, etc. Amounts are in cents, right justified, with leading zeroes entered.
11–12 22–23 33–34 44–45 55–56 66–67 77–78	Number of cells for preceding range. For example, columns 11–12 indicate the number of cells into which the range with the upper limit defined in columns 2–10 will be divided. Amounts are right justified with a leading zero if the number is less than 10.

Process: The record file is read into the computer. The cell for each record is determined based on its transaction amount. The program keeps track of the number of units in each cell, the total dollar amount within each cell, and the sum of squares of dollar amounts within each cell.

Output: For each cell the output shows the dollar range, the number of items, the total dollar amount, and the sum of squares of dollars. Positive amounts, negative amounts, and zero amounts are shown separately.

Example: In the example shown on the following page, the auditor wanted to plan a sample selection from a file of 12222 accounts receivable. The client's file was run against a program which rearranged the record fields to provide a magnetic tape with the amount of each receivable in the first nine positions of each record, the sign in the tenth position and data in which the auditor was interested in positions 11 through 80. The auditor prepared a specification card to obtain statistical data on accounts receivable with debit balances in 10 cells of $100 each in the range $.01 to $1000 and three cells of $500 each in the range $1000.01 to $2,500.00. He prepared another specification card to obtain statistical data on accounts receivable with credit balances in three cells of $50 each in the range $.01 to $150. These cards were coded as follows, beginning with column 1 in each case.

```
PØØØ1ØØØØØ1ØØØØ25ØØØØØ3
NØØØØ15ØØØØ3
```

If there had been any zero balances, positive balances over $2,500 or negative balances over $150, related statistical data would have been shown in separate cells in the output.

RANGE FROM	RANGE TO	NUMBER OF ITEMS	TOTAL DOLLARS	SUM OF SQ. OF DOLLARS
.01	100.00	3730	169,381.73	10,828,099.2933
100.01	200.00	2458	368,712.30	57,412,642.5286
200.01	300.00	2252	566,434.30	144,370,966.4770
300.01	400.00	1613	559,587.29	195,500,317.3173
400.01	500.00	883	394,202.09	176,701,435.8883
500.01	600.00	470	255,140.53	138,883,731.4495
600.01	700.00	217	139,673.40	90,083,519.8742
700.01	800.00	110	81,781.89	60,885,374.1529
800.01	900.00	61	51,472.38	43,483,731.1936
900.01	1,000.00	34	32,024.68	30,193,418.5118
1,000.01	1,500.00	35	41,410.60	49,764,803.2292
1,500.01	2,000.00	2	3,597.59	6,473,957.2045
2,000.01	2,500.00	1	2,361.09	5,574,745.9881
ALL-POSITIVE AMOUNTS		11866	2,665,779.87	1,010,156,743.1083

RANGE FROM	RANGE TO	NUMBER OF ITEMS	TOTAL DOLLARS	SUM OF SQ. OF DOLLARS
.01-	50.00-	350	2,709.07-	47,936.9371
50.01-	100.00-	5	347.22-	25.590.0198
100.01-	150.00-	1	142.51-	20,309.1001
ALL-NEGATIVE-AMOUNTS		356	3,198.80-	93,836.0570
ALL ZERO AMOUNTS		0	.00	
GRAND-TOTAL		12222	2,662,581.07	1,010,250,579.1653

PLAN1

Purpose: To obtain the planned precision and reliability corresponding to specified risks and material amount.

Input: In response to messages printed at the computer terminal, the user designates whether the positive or negative approach is being used, the beta sampling risk, the alpha risk of overauditing, and the material amount.

Process: The program uses the appropriate formula described in chapter 3 to determine the planned precision and reliability. The beta sampling risk must be one of the following: .005, .01, .025, .05, .075, .1, .15, .2, .25, .3, .35, .4, .45, .50. The alpha risk of overauditing must be one of the following: .01, .02, .05, .10, .15, .20, .30, .40, or .50.

Output: The program prints the planned precision (A) together with the appropriate reliability level. The reliability corresponds to interval reliability (R) when the positive approach is used and one-sided reliability (R_1) when the negative approach is used. In the latter case, the result is also given for an interval reliability to facilitate input to other programs.

Example: Two examples of the operation of PLAN1 are shown on the next page. In both cases, the auditor specifies a .05 sampling risk, a .1 risk of overauditing and a $100,000 material amount. The positive approach is specified in the first example and the negative approach in the second example. As can be seen, the calculated precision and reliability are the same for these two examples because the sampling risk is exactly one-half the risk of overauditing.

PLAN2

Purpose: To determine the location of stratum boundaries, the sample size necessary to achieve the planned precision at a specified reliability, and the allocation of the sample size to the strata.

Input: Before running the program, the user enters the number of ranges of data together with the lower limit of the range of smallest amounts. For each range in ascending dollar amount, the following are entered: upper limit, number of items, total dollar amount, and the sum of squares of dollar amounts. This information will normally be obtained from CPA1.

In response to messages printed at the computer terminal the user specifies the desired precision, reliability, number of strata, how the stratum boundaries are to be determined, how the sample is to be allocated to the strata, and whether the sample size is to be determined using the standard deviation of recorded amounts or the approximate expected standard deviation of difference amounts.

```
>RUN

14:23   DEC  9  /PLAN1

IS YOUR APPROACH 1-POSITIVE OR 2-NEGATIVE
1 OR 2  ?1

BETA SAMPLING RISK =  ?.05

ALPHA RISK OF OVERAUDITING =  ?.1

MATERIAL AMOUNT =   ?100000


WITH POSITIVE APPROACH
DESIRED PRECISION (A) =    50000.0
INTERVAL RELIABILITY (R) =   90    PER CENT
INTERVAL RELIABILITY IS REQUIRED AS INPUT TO PROGRAMS FOR
EVALUATION OF SAMPLE RESULTS.

     620 HALT
>
```

```
>RUN

14:24   DEC  9  /PLAN1

IS YOUR APPROACH 1-POSITIVE OR 2-NEGATIVE
1 OR 2  ?2

BETA SAMPLING RISK =  ?.05

ALPHA RISK OF OVERAUDITING =  ?.1

MATERIAL AMOUNT =   ?100000


WITH NEGATIVE APPROACH
DESIRED PRECISION (A) =    50000.0
ONE-SIDED RELIABILITY (R1) =   95    PER CENT
INTERVAL RELIABILITY (R) =   90    PER CENT
INTERVAL RELIABILITY IS REQUIRED AS INPUT TO PROGRAMS FOR
EVALUATION OF SAMPLE RESULTS.

     620 HALT
>
```

Stratum boundaries may be determined using either the square root of the cumulative frequency method or equal dollar amount rule. The allocation may be either the Neyman method or in proportion to the total dollar amount.

Process: The program assigns stratum boundaries beginning with the lowest dollar value range. As each range is read in, the program includes it in the stratum being established if (1) the number of ranges left at least equals the number of strata left and (2) the amount (dollars or square root of the frequency) already assigned to the current stratum plus one-half the amount in the current range does not exceed the average total amount in the current and higher valued strata.

After stratum boundaries are assigned, the program calculates the standard deviation of recorded amounts within each stratum. If the user specified the mean estimate, the program uses a sample size formula corresponding to the stratified mean estimate. The formula used depends upon whether the user specified Neyman allocation or allocation in proportion to the total dollar amount. If the user specified the difference estimate, the approximate standard deviation of the stratum differences is calculated, as described in chapter 6, and the appropriate sample size formula used.

The program identifies high dollar ranges for 100 percent examination if (1) this will reduce the total sample size or (2) the calculated stratum sample size exceeds the stratum population size.

Output: To help the user verify the accuracy of the data entered in the program, the program prints the lower and upper limits of each range of data, the number of items in each range, their total dollar amount, and their sum of squares. After the remaining input is typed in, the program prints the upper limit for each stratum, the total number in the stratum, the calculated sample size for each stratum, and the total dollar amount.

Example: The following two pages illustrate the use of PLAN2 to obtain a sample plan based on the output of the CPA1 and PLAN1 programs illustrated previously. Since the sums of squares are so large, the "E format" is used to enter the amounts in thousands. The number after E indicates how many places the decimal point has been shifted to the left. In both examples, $50,000 precision with 90 percent two-sided reliability and three sample strata are desired.

In the first example, the auditor specifies determination of stratum boundaries by the square root of the cumulative frequency rule, mean estimation, and allocation of the sample among strata by the Neyman method. The printout indicates that a total of 736 items should be selected for examination. These consist of 603 items divided among three sample strata plus all 133 items over $800.

In the second example, the auditor specifies the determination of stratum boundaries by the square root of the cumulative frequency rule and difference estimation with an estimated .07 of items in error. The printout indicates that a total of 582 items should be selected for examination. These consist of 581 items divided among three sample strata plus the single item over $2,000.

```
1 DATA16,-150
10 DATA-100,1,-142.51,20309
20 DATA-50,5,-347.22,25590
30 DATA0,350,-2709.07,47937
40 DATA100,3730,169381.73,10828E3
50 DATA200,2458,368712.3,57412E3
55 DATA300,2252,566434.3,144371E3
60 DATA400,1613,559587.29,195500E3
70 DATA500,883,394202.09,176701E3
80 DATA600,470,255140.53,138884E3
90 DATA700,217,139673.4,90084E3
100 DATA800,110,81781.89,60885E3
110 DATA900,61,51472.38,43484E3
120 DATA1000,34,32024.68,30193E3
130 DATA1500,35,41410.6,49765E3
140 DATA2000,2,3597.59,6474E3
150 DATA2500,1,2361.09,5575E3
>RUN
14:38          /PLAN2
```

FILE PROFILE DATA

LOWER LIMIT	UPPER LIMIT	NUMBER OF ITEMS	DOLLAR AMOUNT	SUM OF SQUARES
-150	-100	1	-142.510	20309
-99.9900	-50	5	-347.220	25590
-49.9900	0	350	-2709.07	47937
1.00000E-02	100	3730	169382.	10828000
100.010	200	2458	368712.	57412000
200.010	300	2252	566434.	144371000
300.010	400	1613	559587.	195500000
400.010	500	883	394202.	176701000
500.010	600	470	255141.	138884000
600.010	700	217	139673.	90084000
700.010	300	110	81781.9	60885000
800.010	900	61	51472.4	43484000
900.010	1000	34	32024.7	30193000
1000.01	1500	35	41410.6	49765000
1500.01	2000	2	3597.59	6474000
2000.01	2500	1	2361.09	5575000
TOTAL DOLLAR AMOUNT			2.66258E+06	

DESIRED PRECISION= ?50000

RELIABILITY PERCENTAGE= ?90

1 OR 2 SIDED ?2

DESIRED NUMBER OF SAMPLE STRATA= ?3

STRATUM BOUNDARIES TO BE DETERMINED BY THE CUM F RULE (1)
OR BY THE EQUAL DOLLAR VALUE RULE (2)
1 OR 2 ?1

MEAN ESTIMATE (1) OR DIFFERENCE ESTIMATE (2)?
1 OR 2 ?1

ALLOCATION OF SAMPLE AMONG STRATA BY NEYMAN FORMULA (1)
OR IN PROPORTION TO TOTAL DOLLAR AMOUNT (2)?
1 OR 2 ?1

STRATUM NUMBER	UPPER LIMIT	TOTAL NR IN STRATUM	SAMPLE SIZE	TOTAL $ IN STRATUM
1	100	4086	102	166183.
2	300	4710	215	935147.
3	800	3293	286	1.43039E+06
	2500	133	133	130866.

```
1 DATA16,-150
10 DATA-100,1,-142.51,20309
20 DATA-50,5,-347.22,25590
30 DATA0,350,-2709.07,47937
40 DATA100,3730,169381.73,10828E3
50 DATA200,2458,368712.3,57412E3
55 DATA300,2252,566434.3,144371E3
60 DATA400,1613,559587.29,195500E3
70 DATA500,883,394202.09,176701E3
80 DATA600,470,255140.53,138884E3
90 DATA700,217,139673.4,90084E3
100 DATA800,110,81781.89,60885E3
110 DATA900,61,51472.38,43484E3
120 DATA1000,34,32024.68,30193E3
130 DATA1500,35,41410.6,49765E3
140 DATA2000,2,3597.59,6474E3
150 DATA2500,1,2361.09,5575E3
>RUN
14:44          /PLAN2
```

FILE PROFILE DATA

LOWER LIMIT	UPPER LIMIT	NUMBER OF ITEMS	DOLLAR AMOUNT	SUM OF SQUARES
-150	-100	1	-142.510	20309
-99.9900	-50	5	-347.220	25590
-49.9900	0	350	-2709.07	47937
1.00000E-02	100	3730	169382.	10828000
100.010	200	2458	368712.	57412000
200.010	300	2252	566434.	144371000
300.010	400	1613	559587.	195500000
400.010	500	883	394202.	176701000
500.010	600	470	255141.	138884000
600.010	700	217	139673.	90084000
700.010	800	110	81781.9	60885000
800.010	900	61	51472.4	43484000
900.010	1000	34	32024.7	30193000
1000.01	1500	35	41410.6	49765000
1500.01	2000	2	3597.59	6474000
2000.01	2500	1	2361.09	5575000
TOTAL DOLLAR AMOUNT			2.66258E+06	

```
DESIRED PRECISION=  ?50000

RELIABILITY PERCENTAGE=   ?90

1 OR 2 SIDED  ?2

DESIRED NUMBER OF SAMPLE STRATA=  ?3

STRATUM BOUNDARIES TO BE DETERMINED BY THE CUM F RULE (1)
OR BY THE EQUAL DOLLAR VALUE RULE (2)
1 OR 2  ?2

MEAN ESTIMATE (1) OR DIFFERENCE ESTIMATE (2)?
1 OR 2  ?2

ESTIMATED PROPORTION OF ITEMS IN ERROR=   ?.07
```

STRATUM NUMBER	UPPER LIMIT	TOTAL NR IN STRATUM	SAMPLE SIZE	TOTAL $ IN STRATUM
1	300	8796	271	1.10133E+06
2	500	2496	188	953789.
3	2000	929	122	605101.
	2500	1	1	2361.09

CPA2

Purpose: To select a stratified random sample from a population of recorded amounts.

Input: Input consists of a record file, in the format described for CPA1, and specification cards. The format of the first specification card is:

Card Column(s)	Description
1–10	A 10-digit random number
13–14	Desired number of sample strata. This amount must be in the range 1 to 50. It must be right justified with a zero in column 13 if the number is less than 10.

An additional specification card is required for each stratum. If the total number of items in the stratum is known, the following format should be used:

Card Column(s)	Description
1–9	Upper stratum limit (largest positive or smallest negative value in the stratum range) in cents, right justified with leading zeroes entered.
10	Blank for a positive limit, or a minus sign (−) for a negative limit.
12–18	Total number of items in stratum, right justified with leading zeroes entered.
20–24	Stratum sample size, right justified with leading zeroes entered.

If the exact number of items in the stratum is not known, the following format should be used:

Card Column(s)	Description
1–9	Upper stratum limit (largest positive or smallest negative value in the stratum range) in cents, right justified with leading zeroes entered.
10	Blank for a positive limit or a minus sign (−) for a negative limit.
25–30	Sampling fraction, expressed as a ratio with an implied decimal before column 25.

Process: As each record is read into the computer, the program determines the stratum to which it belongs and generates a random number between 0 and 1. The record is included in the sample if the random number is less than the current sampling fraction. For those strata defined by a specification card in the first format, the current sampling fraction equals the ratio of the number of sample items remaining to be selected divided by the number of stratum items remaining to be considered.

Output: Each record in the sample is printed in the order and format in which it appears in the record file. The stratum number to which each record belongs is printed to the left of the record. This listing is followed by a summary which shows the following information for each stratum and for all negative, positive, and zero amounts:

Total number in the population
Total dollar amount in the population
Population standard deviation
Sample size
Total dollar amount in sample
Sum of squares of dollar amounts in sample

Example: The examples on the following three pages illustrate the use of CPA2 to select a sample according to the plan obtained from PLAN2 previously described. The specification cards were coded as follows:

```
9816232823  Ø3
ØØØØ3ØØØØ  ØØØ8796 ØØ261
ØØØØ5ØØØØ  ØØØ2496 ØØ188
ØØØ2ØØØØØ  ØØØØ929 ØØ122
```

Only one of 11 pages in the output which list records selected for the sample is displayed here. The portion of the output under "Description" shows the data in the record file after the transaction amount. In this case, the data consist of the customer account number, credit line, opening balance service charge, current balance cash advance, date of last payment, and number of months delinquent.

STRATIFIED SAMPLE

STRATUM	AMOUNT	DESCRIPTION					
1	256.25	2701026936	600.00	7.34		012872	1
3	559.12	2701028395		9.44		033072	
2	384.63	2701028403	400.00	3.75	100.00	041272	
2	334.40	2701031233	300.00	5.04		031572	7
2	314.45	2701031761	500.00	2.12	191.97	041172	
1	35.00-	2701032496		35.00-		011271	
3	519.41	2701034369	500.00	7.54		031572	
3	550.46	2701034641	500.00	7.97		032772	
2	478.24	2701035051	500.00	7.83		032472	
1	95.45	2701036737	400.00	2.12		030872	
2	349.19	2701037495	500.00	4.70	42.58	031472	
1	15.60	2701037610	300.00	.35		031372	
2	470.46	2701038741	700.00	11.65	252.28	032872	
3	607.71	2701038964	600.00	7.05		032172	
2	366.55	2701041117	400.00		359.69	031372	
2	351.73	2701044665	600.00	5.72		033172	
3	559.88	2701048971	650.00	8.49		031572	
4	2,361.09	2701049227	600.00	18.70		020872	1
1	9.37	2701050274	600.00	1.00-		113071	
1	11.71	2701050704	500.00			032072	
3	715.37	2701050910	750.00	1.34	566.64	032372	
1	8.64-	2801002050	300.00	8.64-		042171	
1	48.12	2801002167	400.00	.84		032772	
2	466.27	2801002605	500.00		458.02	032072	
1	214.66	2801003694		3.34		032772	
2	320.31	2801004999	500.00	4.84		033172	
2	491.03	2801006325	700.00	6.54		031672	

3	900.83	2801006408	850.00	10.94	193.01	032372	1
1	113.96	2801008552	400.00	1.80		040772	
1	11.59	2801011630	400.00			010671	
2	438.21	2801013594	300.00	6.26		032972	2
1	10.39	2801017496	500.00	.39		031472	
1	6.04	2801021456	1000.00			032272	
3	522.23	2801022835	600.00	10.53	128.69	040772	
3	545.04	2801024393	500.00	9.37		031072	
2	447.80	2801024435	500.00	6.06		030872	
1	24.96	2801024609	500.00			032072	
1	24.80	2801024948	500.00	.83		032972	
2	441.70	2801026331	500.00	1.15	300.00	041172	
1	1.00-	2801026505	400.00	1.00-		000000	
1	161.35	2801026760	400.00	2.66		040472	
3	535.22	2801027404	500.00	16.97		040472	2
1	255.89	2801028824	500.00	4.17		040472	
2	461.53	2801029848	500.00	11.54		040772	
2	390.15	2801030143	400.00	6.61		031772	
1	63.78	2801031133	600.00	4.97		033072	
2	386.35	2801032040	600.00	3.29	180.74	041272	
3	518.36	2801033568	500.00	7.76		041272	
1	150.69	2801036033	300.00	2.70	65.19	020472	1
2	302.78	2801038112	500.00	4.99		041272	

CPA2 (concluded)

STRATIFIED SAMPLE

	TOTAL NR IN POPULATION	TOTAL $ IN POPULATION	POPULATION STD DEV	SAMPLE SIZE	TOTAL $ IN SAMPLE	SAMPLE SUM OF SQUARES
STRATUM 1						
NEGATIVE	356	3,198.80-	13.5222	13	201.35-	11793.7383
POSITIVE	8440	1,104,528.33	89.8024	258	35,009.29	6783830.9367
TOTAL	8796	1,101,329.53	92.2229	271	34,807.94	6795624.6750
STRATUM 2	2496	953,789.38	55.6612	188	72,689.35	28755666.8165
STRATUM 3	929	605,101.07	166.1249	122	80,224.85	57403860.8301
STRATUM 4	1	2,361.09		1	2,361.09	5574745.9881
ALL STRATA						
NEGATIVE	356	3,198.80-	13.5222	13	201.35-	11793.7383
ZERO						
POSITIVE	11866	2,665,779.87	186.1710	569	190,284.58	98518104.5714
TOTAL	12222	2,662,581.07	187.6142	582	190,083.23	98529898.3097

MPUDIF

Purpose: To calculate the estimated total amount of a population together with the achieved precision at a specified interval reliability using either the stratified mean estimator or the stratified difference estimator. The total amount corresponds to the estimated audited amount if the mean estimator is used and to the estimated difference if the difference estimator is used.

Input: The user enters the interval reliability as a percentage and the number of strata. For each stratum the user enters the population size, the sample size, and the number of non-zero amounts in the sample, followed by the audited amounts or the non-zero differences that were determined.

Process: The program uses the appropriate formula from chapter 6 to compute the estimate of the population total. The standard error is likewise calculated using the formula from chapter 6. The precision of the estimate is calculated using a reliability factor from Student's *t*-distribution. The degrees of freedom are calculated from the formula given in Cochran [5], p. 95, based on the number of observed non-zero differences for the difference estimator.

Output: The output shows a summary of the input, a summary of results, and the statistical details.

The summary of input shows the population size of each stratum, the stratum sample size, the sample total amount within each stratum, and the number of non-zero amounts within each stratum.

The summary of results shows the estimate of the population total, the achieved precision at the specified interval reliability, and the lower and upper precision limits.

The statistical details include, for each stratum, the sample mean, the sample estimate of the total stratum amount, the estimated standard deviation, and the estimated standard error. Also shown are the calculated degrees of freedom and the corresponding *t*-factor used in calculating the achieved precision.

Example: The following page shows the use of this program to evaluate the results of auditing the sample selected by the CPA2 program. In this example, the evaluation is based on the observed differences, and the estimate of the population total refers to the estimated difference between the audited and recorded amounts.

```
1  DATA 90, 3
10 DATA 8976, 271, 10
12 DATA  -23.78,  -12.14,     -.32,  -22.51, -173.04, -160.45,  -35.69
14 DATA  -87.98,    -6.33,  -80.62
20 DATA 2496, 188, 11
22 DATA  -62.65, -146.93,  -73.51,   89.20, -151.46, -311.46, -270.24
24 DATA -328.90, -347.96, -165.39, -320.43
30 DATA 929, 122, 11
32 DATA -245.67,  242.10, -537.28, -370.99, -752.49, -244.31, -494.03
34 DATA -515.91,   57.72, -102.76, -110.13
>RUN
12:37   DEC 27  /MPUDIF

  SUMMARY OF INPUT

STRATUM        TOTAL NR IN    SAMPLE        TOTAL $ IN     NR OF NON-
NUMBER           STRATUM       SIZE           SAMPLE       ZERO AMOUNTS
 1              8976          271           -602.860        10
 2              2496          188           -2089.73        11
 3              929           122           -3073.75        11

  SUMMARY OF RESULTS

ESTIMATE OF POPULATION TOTAL                -71118.1
PRECISION AT   90   %
  INTERVAL RELIABILITY                       26880
  LOWER PRECISION LIMIT                     -97998.1
  UPPER PRECISION LIMIT                     -44238.1

  STATISTICAL DETAILS

STRATUM        SAMPLE         SAMPLE        ESTIMATED      ESTIMATED
NUMBER          MEAN         ESTIMATE        STD DEV       STD ERROR
 1             -2.22458      -19967.8        16.2298        8714.72
 2             -11.1156      -27744.5        55.1082        9646.67
 3             -25.1947      -23405.9        116.065        9098.43
ALL                          -71118.1                       15867.8

T FACTOR FOR   31.9050     DEGREES OF FREEDOM               1.69400

PRECISION EQUALS ESTIMATED STD ERROR TIMES T FACTOR         26880


    1770 HALT
>
```

COMRAT

Purpose: To calculate the estimated total audited amount of a population together with the achieved precision at a specified interval reliability using the combined ratio estimator.

Input: The user enters the interval reliability as a percentage, the data format (1 or 2), and the number of strata.

Using data format 1, for each stratum the user enters the population size, the total recorded amount, and the sample size followed by the recorded amount of each sample unit together with the observed difference (negative for overstatement, positive for understatement, zero if correct).

Using data format 2, for each stratum the user enters the population size, the total recorded amount, the sample size, the recorded amount in the sample, the sum of squares of recorded amounts in the sample, and the number of non-zero differences followed by the recorded amount and the observed difference for each observation with a non-zero difference.

Process: The program uses the appropriate formulas from chapter 6 to compute the combined ratio estimate of the population audited amount and the standard error. The precision is calculated using a reliability factor from Student's t-distribution. The degrees of freedom are calculated from the formula given in Cochran [5], p. 95, based on the number of observed non-zero differences.

Output: The output shows a summary of the input, a summary of the results, and the statistical details.

The summary of the input shows, for each stratum, the population size, the total recorded amount, the sample size, and the total recorded amount within the sample. In addition, for each stratum, the printout shows the average recorded amount for the population and for the sample, the number of observed differences, and the dollar amount of these differences.

The summary of results shows the total recorded amount, the combined ratio estimate of the total difference, the estimate of the total audited amount (the sum of the first two), the achieved precision of the estimate at the specified interval reliability, and the lower and upper precision limits.

The statistical details include, for each stratum, the estimated standard deviation, the estimated standard error, and the sample average recorded amount and sample average audited amount, each multiplied times the stratum sample size. There follows a step-by-step derivation of the combined ratio estimate, the calculated degrees of freedom, the corresponding t-factor, and the calculated precision of the estimate.

Example: The following page shows the use of this program to evaluate the results of auditing the sample selected by the CPA2 program. Because the combined ratio estimate is only valid for positive recorded amounts, the evaluation is limited to the 568 sample items with positive recorded amounts.

```
1  DATA 90, 2, 3
10 DATA 8440, 1104528.33, 258, 35009.29, 6783831, 10
12 DATA  89.43, -23.78,   4.25, -12.14,  32.65,   -.32, 102.36, -22.51
14 DATA 173.04,-173.04, 178.56,-160.45, 102.77, -35.69,  98.12, -87.98
16 DATA  42.61,  -6.33,  80.62, -80.62
20 DATA 2496, 953789.38, 188, 72689.35, 28755667, 11
22 DATA 345.65, -62.65, 374.14,-146.93, 336.06, -73.51, 460.00,  89.20
24 DATA 385.67,-151.46, 311.46,-311.46, 460.33,-270.24, 383.67,-328.90
26 DATA 301.12,-347.96, 320.31,-165.39, 320.43,-320.43
30 DATA 929, 605101.07, 122, 80224.85, 57403861, 11
32 DATA 714.09,-245.67, 636.54, 242.10, 537.28,-537.28, 638.11,-370.99
34 DATA 752.49,-752.49,  28.34,-244.31, 525.10,-494.03, 729.59,-515.91
36 DATA 730.53,  57.72, 566.95,-102.76, 500.66,-110.13
>RUN
14:43    DEC 28   /COMRAT

  SUMMARY OF INPUT

STRATUM         TOTAL NR IN    TOTAL $ IN       SAMPLE          TOTAL $ IN
NUMBER            STRATUM        STRATUM         SIZE             SAMPLE
 1               8440           1.10453E+06     258              35009.3
 2               2496           953789.         188              72689.4
 3               929            605101.         122              80224.9

STRATUM         AVERAGE RECORDED AMOUNT          DIFFERENCES FOUND
NUMBER           STRATUM        SAMPLE          NUMBER          DOLLARS
 1               130.868        135.695         10              -602.860
 2               382.127        386.645         11              -2089.73
 3               651.347        657.581         11              -3073.75

  SUMMARY OF RESULTS

TOTAL RECORDED AMOUNT                           2.66342E+06
ESTIMATE OF TOTAL DIFFERENCE                   -69366.3
ESTIMATE OF POPULATION TOTAL                    2.59405E+06
PRECISION AT   90   %
  INTERVAL RELIABILITY                          26924.5
  LOWER PRECISION LIMIT                         2.56713E+06
  UPPER PRECISION LIMIT                         2.62098E+06

  STATISTICAL DETAILS

STRATUM         ESTIMATED      ESTIMATED       EXTENDED SAMPLE AVERAGES
NUMBER           STD DEV       STD ERROR       RECORDED AMT  AUDITED AMT
 1               16.8286        8706.40         1.14527E+06   1.12554E+06
 2               55.3444        9688.02         965067.       937323.
 3               116.818        9157.45         610893.       587487.
ALL                             15922.3         2.72122E+06   2.65035E+06

RATIO OF EXT AVE AUDITED TO RECORDED AMT        .973956
MULTIPLY BY TOTAL RECORDED AMOUNT               2.66342E+06
ESTIMATE OF POPULATION TOTAL                    2.59405E+06

T FACTOR FOR   33.8830      DEGREES OF FREEDOM            1.69100
PRECISION EQUALS ESTIMATED STD ERROR TIMES T FACTOR       26924.5


    2260 HALT
>
```

COMREG

Purpose: To calculate the estimated total audited amount of a population together with the achieved precision at a specified interval reliability using the combined regression estimator.

Input: The user enters the interval reliability as a percentage, the data format (1 or 2), and the number of strata.

Using data format 1, for each stratum the user enters the population size, the total recorded amount, and the sample size followed by the recorded amount of each sample unit together with the observed difference (negative for overstatement, positive for understatement, zero if correct).

Using data format 2, for each stratum the user enters the population size, the total recorded amount, the sample size, the recorded amount in the sample, the sum of squares of recorded amounts in the sample, and the number of non-zero differences followed by the recorded amount and the observed difference for each observation with a non-zero difference.

Process: The program uses the appropriate formulas from chapter 6 to compute the combined regression estimate of the population audited amount and the standard error. The precision is calculated using a reliability factor from Student's *t*-distribution. The degrees of freedom are calculated from the formula given in Cochran [5], p. 95, based on the number of observed non-zero differences.

Output: The output shows a summary of the input, a summary of the results, and the statistical details.

The summary of the input shows, for each stratum, the population size, the total recorded amount, the sample size, and the total recorded amount within the sample. In addition, for each stratum, the printout shows the average recorded amount for the population and for the sample, the number of observed differences, and the dollar amount of these differences.

The summary of results shows the total recorded amount, the combined regression estimate of the total difference, the estimate of the total audited amount (the sum of the first two), the achieved precision of the estimate at the specified interval reliability, and the lower and upper precision limits.

The statistical details include, for each stratum, the estimated standard deviation, the estimated standard error, and the sample average recorded amount and sample average audited amount, each multiplied times the stratum sample size. There follows a step-by-step derivation of the combined regression estimate, the calculated degrees of freedom, the corresponding *t*-factor, and the calculated precision of the estimate.

Example: The following page shows the use of this program to evaluate the results of auditing the sample selected by the CPA2 program.

```
1  DATA 90, 2, 3
10 DATA 8796, 1101329.53, 271, 34807.94, 6795625, 10
12 DATA  89.43, -23.78,    4.25, -12.14,  32.65,     -.32, 102.36, -22.51
14 DATA 173.04,-173.04, 178.56,-160.45, 102.77, -35.69,  98.12, -87.98
16 DATA  42.61,  -6.33,  80.62, -80.62
20 DATA 2496, 953789.38, 188, 72689.35, 28755667, 11
22 DATA 345.65, -62.65, 374.14,-146.93, 336.06, -73.51, 460.00,  89.20
24 DATA 385.67,-151.46, 311.46,-311.46, 460.33,-270.24, 383.67,-328.90
26 DATA 301.12,-347.96, 320.31,-165.39, 320.43,-320.43
30 DATA 929, 605101.07, 122, 80224.85, 57403861, 11
32 DATA 714.09,-245.67, 636.54, 242.10, 537.28,-537.28, 638.11,-370.99
34 DATA 752.49,-752.49,  28.34,-244.31, 525.10,-494.03, 729.59,-515.91
36 DATA 730.53,  57.72, 566.95,-102.76, 500.66,-110.13
>RUN
11:59   DEC 27  /COMREG

  SUMMARY OF INPUT

STRATUM        TOTAL NR IN    TOTAL $ IN     SAMPLE         TOTAL $ IN
NUMBER           STRATUM        STRATUM       SIZE            SAMPLE
 1             8796           1.10133E+06    271             34807.9
 2             2496           953789.        188             72689.4
 3             929            605101.        122             80224.9

STRATUM        AVERAGE RECORDED AMOUNT       DIFFERENCES FOUND
NUMBER           STRATUM      SAMPLE         NUMBER         DOLLARS
 1               125.208      128.443        10             -602.860
 2               382.127      386.645        11             -2089.73
 3               651.347      657.581        11             -3073.75

  SUMMARY OF RESULTS

TOTAL RECORDED AMOUNT                         2.66022E+06
ESTIMATE OF TOTAL DIFFERENCE                 -71024.1
ESTIMATE OF POPULATION TOTAL                  2.58920E+06
PRECISION AT    90    %
  INTERVAL RELIABILITY                        26819.4
  LOWER PRECISION LIMIT                       2.56238E+06
  UPPER PRECISION LIMIT                       2.61602E+06

  STATISTICAL DETAILS

STRATUM        ESTIMATED      ESTIMATED     EXTENDED SAMPLE AVERAGES
NUMBER           STD DEV      STD ERROR     RECORDED AMT   AUDITED AMT
 1               16.2792       8563.25      1.12978E+06    1.11021E+06
 2               55.2044       9663.50      965067.        937323.
 3               116.462       9129.52      610893.        587487.
ALL                           15813.3       2.70574E+06    2.63502E+06

TOTAL RECORDED AMOUNT                         2.66022E+06
LESS EXTENDED SAMPLE AVE RECORDED AMT         2.70574E+06
DIFFERENCE                                   -45520.6
TIMES COMBINED B FACTOR                       1.00673
PRODUCT                                      -45827.0
PLUS EXTENDED SAMPLE AVE AUDITED AMT          2.63502E+06
ESTIMATE OF POPULATION TOTAL                  2.58920E+06

T FACTOR FOR   30.8750      DEGREES OF FREEDOM             1.69600
PRECISION EQUALS ESTIMATED STD ERROR TIMES T FACTOR        26819.4


    2350 HALT
>
```

SEPRAT

Purpose: To calculate the estimated audited amount of a population together with the achieved precision at a specified interval reliability using the separate ratio estimator.

Input: The user enters the interval reliability as a percentage, the data format (1 ot 2), and the number of strata.

Using data format 1, for each stratum the user enters the population size, the total recorded amount, and the sample size followed by the recorded amount of each sample unit together with the observed difference (negative for overstatement, positive for understatement, zero if correct).

Using data format 2, for each stratum the user enters the population size, the total recorded amount, the sample size, the recorded amount in the sample, the sum of squares of recorded amounts in the sample, and the number of non-zero differences followed by the recorded amount and the observed difference for each observation with a non-zero difference.

Process: The program uses the formulas described in Appendix 7 to compute the separate ratio estimate of the population audited amount and the standard error. The precision is calculated using a reliability factor from Student's *t*-distribution. The degrees of freedom are calculated from the formula given in Cochran [5], p. 95, based on the number of observed non-zero differences.

Output: The output shows a summary of the input, a summary of the results, and the statistical details.

The summary of the input shows, for each stratum, the population size, the total recorded amount, the sample size, and the total recorded amount within the sample. In addition, for each stratum, the printout shows the average recorded amount for the population and for the sample, the number of observed differences, and the dollar amount of these differences.

The summary of results shows the total recorded amount, the separate ratio estimate of the total difference, the estimate of the total audited amount (the sum of the first two), the achieved precision of the estimate at the specified interval reliability, and the lower and upper precision limits.

The statistical details include, for each stratum, the sample ratio, the estimated audited amount, the estimated standard deviation, and the estimated standard error. There follows the calculated degrees of freedom, the corresponding *t*-factor, and the calculated precision of the estimate.

Example: The following page shows the use of this program to evaluate the results of auditing the sample selected by the CPA2 program. Because the separate ratio estimate is only valid when all stratum recorded amounts have the same sign, the evaluation is limited to the 568 sample items with positive recorded amounts.

```
1  DATA 90, 2, 3
10 DATA 8440, 1104528.33, 258, 35009.29, 6783831, 10
12 DATA  89.43, -23.78,   4.25, -12.14,  32.65,   -.32, 102.36, -22.51
14 DATA 173.04,-173.04, 178.56,-160.45, 102.77, -35.69,  98.12, -87.98
16 DATA  42.61,  -6.33,  80.62, -80.62
20 DATA 2496, 953789.38, 188, 72689.35, 28755667, 11
22 DATA 345.65, -62.65, 374.14,-146.93, 336.06, -73.51, 460.00,  89.20
24 DATA 385.67,-151.46, 311.46,-311.46, 460.33,-270.24, 383.67,-328.90
26 DATA 301.12,-347.96, 320.31,-165.39, 320.43,-320.43
30 DATA 929, 605101.07, 122, 80224.85, 57403861, 11
32 DATA 714.09,-245.67, 636.54, 242.10, 537.28,-537.28, 638.11,-370.99
34 DATA 752.49,-752.49,  28.34,-244.31, 525.10,-494.03, 729.59,-515.91
36 DATA 730.53,  57.72, 566.95,-102.76, 500.66,-110.13
>RUN

14:38   DEC 28  /SEPRAT

   SUMMARY OF INPUT

STRATUM        TOTAL NR IN    TOTAL $ IN      SAMPLE       TOTAL $ IN
NUMBER           STRATUM        STRATUM        SIZE          SAMPLE
 1             8440           1.10453E+06     258           35009.3
 2             2496           953789.         188           72689.4
 3             929            605101.         122           80224.9

STRATUM        AVERAGE RECORDED AMOUNT         DIFFERENCES FOUND
NUMBER          STRATUM        SAMPLE        NUMBER        DOLLARS
 1              130.868        135.695        10           -602.860
 2              382.127        386.645        11           -2089.73
 3              651.347        657.581        11           -3073.75

   SUMMARY OF RESULTS

TOTAL RECORDED AMOUNT                          2.66342E+06
ESTIMATE OF TOTAL DIFFERENCE                  -69624.2
ESTIMATE OF POPULATION TOTAL                   2.59379E+06
PRECISION AT    90    %
   INTERVAL RELIABILITY                        26913.3
   LOWER PRECISION LIMIT                       2.56688E+06
   UPPER PRECISION LIMIT                       2.62071E+06

   STATISTICAL DETAILS

STRATUM        SAMPLE         SAMPLE          ESTIMATED      ESTIMATED
NUMBER         RATIO          ESTIMATE        STD DEV        STD ERROR
 1              .982780        1.08551E+06     16.6963        8637.95
 2              .971251        926369.         55.3604        9690.81
 3              .961686        581917.         116.836        9158.83
ALL                            2.59379E+06                    15887.4

T FACTOR FOR   31.8840        DEGREES OF FREEDOM             1,69400

PRECISION EQUALS ESTIMATED STD ERROR TIMES T FACTOR           26913.3


    2220 HALT
>
```

SEPREG

Purpose: To calculate the estimated audited amount of a population together with the achieved precision at a specified interval reliability using the separate regression estimator.

Input: The user enters the interval reliability as a percentage, the data format (1 or 2), and the number of strata.

Using data format 1, for each stratum the user enters the population size, the total recorded amount, and the sample size followed by the recorded amount of each sample unit together with the observed difference (negative for overstatement, positive for understatement, zero for correct).

Using data format 2, for each stratum the user enters the population size, the total recorded amount, the sample size, the recorded amount in the sample, the sum of squares of recorded amounts in the sample, and the number of non-zero differences followed by the recorded amount and the observed difference for each observation with a non-zero difference.

Process: The program uses the formulas described in Appendix 7 to compute the separate regression estimate of the population audited amount and the standard error. The precision is calculated using a reliability factor from Student's *t*-distribution. The degrees of freedom are calculated from the formula given in Cochran [5], p. 95, based on the number of observed non-zero differences.

Output: The output shows a summary of the input, a summary of the results, and the statistical details.

The summary of the input shows, for each stratum, the population size, the total recorded amount, the sample size, and the total recorded amount within the sample. In addition, for each stratum, the printout shows the average recorded amount for the population and for the sample, the number of observed differences, and the dollar amount of these differences.

The summary of results shows the total recorded amount, the separate regression estimate of the total difference, the estimate of the total audited amount (the sum of the first two), the achieved precision of the estimate at the specified interval reliability, and the lower and upper precision limits.

The statistical details include, for each stratum, the sample regression coefficients, the estimated audited amount, the estimated standard deviation, and the estimated standard error. There follows the calculated degrees of freedom, the corresponding *t*-factor, and the calculated precision of the estimate.

Example: The following page shows the use of this program to evaluate the results of auditing the sample selected by the CPA2 program.

```
1  DATA 90, 2, 3
10 DATA 8796, 1101329.53, 271, 34807.94, 6795625, 10
12 DATA  89.43, -23.78,    4.25, -12.14,  32.65,    -.32, 102.36, -22.51
14 DATA 173.04,-173.04, 178.56,-160.45, 102.77, -35.69,  98.12, -87.98
16 DATA  42.61,  -6.33,  80.62, -80.62
20 DATA 2496, 953789.38, 188, 72689.35, 28755667, 11
22 DATA 345.65, -62.65, 374.14,-146.93, 336.06, -73.51, 460.00,  89.20
24 DATA 385.67,-151.46, 311.46,-311.46, 460.33,-270.24, 383.67,-328.90
26 DATA 301.12,-347.96, 320.31,-165.39, 320.43,-320.43
30 DATA 929, 605101.07, 122, 80224.85, 57403861, 11
32 DATA 714.09,-245.67, 636.54, 242.10, 537.28,-537.28, 638.11,-370.99
34 DATA 752.49,-752.49,  28.34,-244.31, 525.10,-494.03, 729.59,-515.91
36 DATA 730.53,  57.72, 566.95,-102.76, 500.66,-110.13
>RUN

11:48    DEC 27   /SEPREG

  SUMMARY OF INPUT

STRATUM          TOTAL NR IN     TOTAL $ IN       SAMPLE          TOTAL $ IN
NUMBER             STRATUM         STRATUM         SIZE             SAMPLE
 1               8796            1.10133E+06      271              34807.9
 2               2496            953789.          188              72689.4
 3               929             605101.          122              80224.9

STRATUM          AVERAGE RECORDED AMOUNT           DIFFERENCES FOUND
NUMBER            STRATUM        SAMPLE          NUMBER           DOLLARS
 1                125.208        128.443          10              -602.860
 2                382.127        386.645          11              -2089.73
 3                651.347        657.581          11              -3073.75

  SUMMARY OF RESULTS

TOTAL RECORDED AMOUNT                             2.66022E+06
ESTIMATE OF TOTAL DIFFERENCE                     -72310.2
ESTIMATE OF POPULATION TOTAL                      2.58791E+06
PRECISION AT    90   %
  INTERVAL RELIABILITY                            26760.6
  LOWER PRECISION LIMIT                           2.56115E+06
  UPPER PRECISION LIMIT                           2.61467E+06

  STATISTICAL DETAILS

STRATUM          ESTIMATED       ESTIMATED        ESTIMATED       ESTIMATED
NUMBER           B FACTOR        POP TOTAL        STD DEV         STD ERROR
 1                .997969         1.08182E+06     16.2588          8552.52
 2                1.12487        924637.          54.7604          9585.78
 3                1.04179        581453.          116.257          9113.48
ALL                              2.58791E+06                       15750.8

T FACTOR FOR    28.9140       DEGREES OF FREEDOM                   1.69900

PRECISION EQUALS ESTIMATED STD ERROR TIMES T FACTOR                26760.6


    2220 HALT

>
```

CPA3

Purpose: To select a probability proportional to size sample from a population of recorded amounts.

Input: Input consists of a record file, in the format described for CPA1, and a specification card. The format of the specification card is:

Card Column(s)	Description
1–6	A 6-digit random number.
8–16	High dollar cutoff (amount *above* which all items are to be selected) in *dollars,* right justified with leading zeroes entered.
17	Sign of records from which selection will be made: blank if positive or a minus sign (−) if negative.
19–30	Total value of all records with the indicated sign which are less than or equal to the high dollar cutoff. This amount is in *cents,* right justified with leading zeroes entered.
32–35	Desired sample size.

Process: The program generates random numbers from 0 to 1 and multiplies them by the total book value. After the number of random numbers generated equals the sample size, the numbers are sorted in ascending order. The first record is then read into the computer. The program reverses its sign if a minus sign has been entered in column 17 of the specification card. If the value of the record exceeds the high dollar cutoff it is printed with an asterisk and a new record is read. If the value is positive and equal to or less than the high dollar cutoff, it is added to an accumulator. If the value of the accumulator is less than the first random number, the next record is read. If it equals or exceeds the random number, it is compared against subsequent random numbers until a random number is found which exceeds it. This record is printed together with the number of random numbers which were exceeded by the cumulative total and the next record is read.

Output: In addition to the records selected for examination, the program prints a summary showing the number and dollar value of (1) items above the high dollar cutoff, (2) items selected for the sample, (3) all positive items, (4) all negative items, and (5) all zero items.

Example: The example on the next two pages illustrates the use of CPA3 to select a probability proportional to size sample from the positive records in the file described under CPA1. The specification card was coded as follows:

783Ø46 ØØØØØ2ØØØ ØØØ266341878 Ø568.

The first page is one of eleven pages in the output which list records selected for examination. The second page summarizes the data in the file and the sample. If a sample of negative records were desired, another run would be required.

OCCUR	VALUE			DESCRIPTION			
	275.13	2701038949	500.00	4.36		040472	
	239.24	2701039194		3.83		040772	
	470.32	2701040739	600.00	2.23	251.80	031472	
	887.81	2701041067	800.00	9.56	295.33	040572	
	259.22	2701041588		3.99		032772	1
	384.64	2701043428	400.00	4.38	50.00	040472	
	260.79	2701045100	500.00	1.88	149.51	040772	
	277.57	2701045266	300.00		272.81	031372	
	443.37	2701045878	600.00	6.59		032272	
	559.32	2701046033	500.00	15.42	37.07	040772	2
	1,125.90	2701046157	750.00	51.68		032772	2
002	813.69	2701047213	1000.00	10.97		031572	
	216.43	2701047320	300.00	.83	159.14	031072	
	282.49	2701048138	600.00	4.15	35.79	033072	
	391.98	2701048583	500.00	4.83		041172	
	270.88	2701048823	300.00	1.89	76.04	040472	
	567.21	2701048856	500.00	15.49		041272	1
*	2,361.09	2701049227	600.00	18.70		020872	1
	475.27	2701050449	500.00	3.12	271.45	040772	

251.29	2701050993	300.00		000000	
108.33	2801000724	400.00	1.36	032272	
872.96	2801000831		13.42	030772	1

SUB-TOTAL
11,794.93

TOTAL VALUE 100% ITEMS 2,361.09	NUMBER 100% ITEMS 1
TOT VAL NON-100% ITEMS 205,703.47	NUMBER NON-100% ITEMS 547
TOTAL VALUE ITEMS 208,064.56	TOTAL NUMBER ITEMS 548
	SAMPLE SIZE NON-100% 568

FILE TOTAL-POSITIVES 2,665,779.87	FILE TOTAL-NEGATIVES 3,198.80	FILE NET TOTAL 2,662,581.07
NUMBER OF POSITIVES 11,866	NUMBER OF NEGATIVES 356	TOTAL NO. OF RECORDS 12,222
		I.D. CRITERIA NOT MET 0
	NUMBER OF ZEROS 0	TOTAL NO. DATA ERRORS 0

PPS

Purpose: To calculate the estimated total audited amount of a population together with the achieved precision at a specified interval reliability using the pps estimator.

Input: The user first enters the interval reliability as a percentage and the number of strata. For each stratum the user then enters the total recorded amount, the sample size, and the number of non-zero differences found followed by the recorded amount and the observed difference for each non-zero difference. (Differences are negative for overstatements and positive for understatements.)

Process: The program uses the formulas described in chapter 6 to compute the pps estimate of the population audited amount and the standard error of the estimate. The precision is calculated using a reliability factor from Student's *t*-distribution. The degrees of freedom are calculated based on the formula given in Cochran [6], p. 95, based on the number of observed non-zero differences.

Output: The output shows a summary of the input, a summary of results, and the statistical details. The summary of input shows the recorded amount, audited amount, and ratio for each sample item with a non-zero difference. For each stratum, the total ratios are divided by the sample size to obtain the mean ratio, and the mean ratio is multiplied by the total recorded amount to obtain the sample estimate of the stratum total.

The summary of results shows the total recorded amount, the pps estimate of the total difference, the estimate of the total audited amount (the sum of the first two), the achieved precision of the estimate, the specified interval reliability, and the lower and upper precision limits.

The statistical details include, for each stratum, the estimated standard deviation and the estimated standard error. There follow the calculated degrees of freedom, the corresponding *t*-factor, and the calculated precision of the sample estimate.

Example: The following page shows the use of this program to evaluate the results of auditing the sample selected by the CPA3 program.

```
1 DATA 90,1
2 DATA 2663418.78,568,30
10 DATA 352.29,-352.29,537.28,-537.28,600.78,-600.78,262.00,-262.00
12 DATA 205.46,-119.04,383.61,-374.68,278.81,-278.81,454.11,-160.49
14 DATA 409.36,-157.50,457.22,-267.87,929.93,-68.31,633.08,80.05
16 DATA 573.50,-573.50,325.16,-142.59,332.59,-200.61,748.04,-748.04
18 DATA 434.24,-434.24,494.86,1.94,399.82,-117.85,449.16,-141.64
20 DATA 151.50,-72.50,227.59,-1.68,627.72,-231.68,273.18,-153.27
22 DATA 443.37,-224.45,270.88,74.74,567.21,-255.49,301.74,-54.81
24 DATA 301.74,-54.81,446.24,-7.46
>RUN

17:07   NOV 29  /PPS

RECORDED AMT  AUDITED AMT    RATIO
 352.290       0              0
 537.280       0              0
 600.780       0              0
 262           0              0
 205.460       86.4200        .420617
 383.610       8.93000        2.32789E-02
 278.810       0              0
 454.110       293.620        .646583
 409.360       251.860        .615253
 457.220       189.350        .414133
 929.930       861.620        .926543
 633.080       713.130        1.12645
 573.500       0              0
 325.160       182.570        .561477
 332.590       131.980        .396825
 748.040       0              0
 434.240       0              0
 494.860       496.800        1.00392
 399.820       281.970        .705242
 449.160       307.520        .684656
 151.500       79             .521452
 227.590       225.910        .992618
 627.720       396.040        .630918
 273.180       119.910        .438941
 443.370       218.920        .493764
 270.880       345.620        1.27592
 567.210       311.720        .549567
 301.740       246.930        .818354
 301.740       246.930        .818354
 446.240       438.780        .983283
OTHER                        538
TOTAL RATIOS                 553.048
DIVIDE BY SAMPLE SIZE        568
MEAN RATIO                   .973676
TIMES TOTAL RECORDED         2.66342E+06

 SUMMARY OF RESULTS

TOTAL RECORDED AMOUNT        2.66342E+06
ESTIMATED TOTAL DIFFERENCE  -70111.0
ESTIMATED POPULATION TOTAL   2.59331E+06
PRECISION AT   90   %
  INTERVAL RELIABILITY       27073.6
  LOWER PRECISION LIMIT      2.56623E+06
  UPPER PRECISION LIMIT      2.62038E+06

  STATISTICAL DETAILS

ESTIMATED STD DEVIATION      .142773

TOTAL ESTIMATED STD ERROR    15953.8

TIMES T FACTOR FOR   30
  DEGREES OF FREEDOM         1.69700

PRECISION OF ESTIMATE        27073.6
```

VAREVA

Purpose: To evaluate the sample estimate of the total audited amount together with the achieved precision to decide whether there might be a material amount of error. The evaluation takes into account the possible difference between planned and achieved precision as well as sensitivity of the estimated standard error.

Input: In response to messages printed at the computer terminal, the user types the beta sampling risk, the material amount, and the following results of the statistical evaluation: total recorded amount (including any stratum sampled 100 percent), estimated total audited amount (including any stratum sampled 100 percent), interval reliability percentage, the achieved precision at the reliability, and the degrees of freedom.

Process: The program first tests for equality between the interval reliability and $(1 - 2\beta)$. If not equal, an adjustment is made to the achieved precision as described in Appendix 5A. The sensitivity of the estimated standard error is tested for underestimation of 10 percent, 20 percent, and 30 percent as described in Appendix 5A.

Output: The program prints one of the following five messages:

1. "Based on the statistical evidence, there may be a material amount of monetary error."
2. "Based on the statistical evidence, there is no material error provided the standard error is closely estimated."
3. "Based on the statistical evidence, there is no material error even if the standard error is underestimated by 10 percent."
4. "Based on the statistical evidence, there is no material error even if the standard error is underestimated by 20 percent."
5. "Based on the statistical evidence, there is no material error even if the standard error is underestimated by 30 percent."

Example: The following page shows the use of VAREVA to help the auditor decide whether there might be a material amount of error based on the pps evaluation of sample results. The total recorded amount and sample estimate of the total audited amount entered into the VAREVA program were obtained from the corresponding amounts shown on the pps printout by adding the $2,361 item in the 100 percent stratum and subtracting $3,199 for the negative items. This assumes that no error was found in the $2,361 item and that the 356 negative items were regarded as a separate population.

```
>RUN

08:47   DEC 28  /VAREVA

   PLANNING DATA

BETA SAMPLING RISK =  ?.05

MATERIAL AMOUNT =   ?100000

  RESULTS OF SAMPLE EVALUATION

TOTAL RECORDED AMOUNT (INCLUDING ANY STRATUM SAMPLED 100%) =  ?2662581

SAMPLE ESTIMATE OF TOTAL AUDITED AMOUNT (INCLUDING ANY STRATUM
SAMPLED 100%) =   ?2592472

INTERVAL RELIABILITY PERCENTAGE =   ?90

PRECISION AT   90   % INTERVAL RELIABILITY =  ?27074



BASED ON THE STATISTICAL EVIDENCE, THERE IS NO MATERIAL
ERROR EVEN IF  THE STANDARD ERROR IS UNDERSTATED
BY   10    PERCENT.

     730 HALT
>
```

10

Problems of Implementation

What does a practice unit need to do in order to use statistical sampling? This question must be answered by any practice unit planning to introduce the use of statistical techniques into its audit function. A variety of answers is possible; the right answer for one practice unit may not be right for another. Each practice unit should develop an implementation strategy that is consistent with its audit philosophy.

The problems of implementation have been grouped within this chapter into four major areas—written policies, training, working paper documentation, and review procedures. The discussion in each area focuses on the important questions that need to be resolved. Some concern problems associated with introducing a new technology into the audit practice, while others pertain to the on-going operation and quality control of the auditing function.

In the final section of the chapter, is a brief discussion of a few problems that practice units have encountered in implementing statistical sampling in their audit practice.

Written Policy

The objective of a written policy concerning statistical sampling is to provide guidance to the audit team. The policy statement should address such issues as when statistical sampling should be considered, who is authorized to use it, and how the sample is to be planned, executed, evaluated, and controlled. The policy might also specify any required documentation.

When to Use Statistical Sampling

The policy statement concerning when statistical sampling is to be used may be either very general or very specific. For example, a general statement that statistical sampling should be considered for any test of details when less than 100 percent of the population items is to be examined allows the audit team complete discretion in deciding whether to use statistical sampling. At the other extreme, a statement that statistical sampling is required in any test of details when less than 100 percent of the population items is to be examined, allows the audit team no discretion. A policy statement somewhere between these two extremes might specify some areas where statistical sampling is required and permit discretion in others.

If the practice unit decides that the audit team should exercise either some or complete discretion in deciding whether to use statistical sampling, it would be helpful to provide some guidelines for making that determination. Useful guidelines should discuss the rationale for selecting statistical sampling instead of a judgmental sample. Since the basic purpose of a statistical sample is to control sampling risk, the rationale should include an explanation of when sampling risk may or may not be important. For example, if there is small risk of a material error, sampling risk may not be important. If an audit procedure has little chance of detecting a material error, the additional sampling risk may be unimportant. If there are several sources of evidence, the sampling risk of any single source may not be important.

Guidelines that give reasons for using statistical sampling as well as for using judgment sampling seem more appropriate than those that attempt to spell out each audit situation where statistical sampling should or should not be used.

Who Will Use Statistical Sampling

A general policy statement concerning who is authorized to use statistical sampling should reflect the requirement of the first general standard pertaining to technical training and proficiency. Specifically, a person who uses statistical sampling in auditing should have training and proficiency in the following areas:

1. Benefits and limitations of the alternative sampling methods permitted by the practice unit.
2. Proper application of the more commonly used methods.
3. Relation of sample results to audit conclusions.
4. Documentation of applications.

The need for a written policy pertaining to who may use statistical sampling is most obvious in the beginning stages of its introduction into an audit practice. Typically, at that point few persons within the practice unit are trained, and, consequently, a policy statement may be necessary to restrict the use to those who have the requisite training and proficiency.

The policy statement might also specify when statistical audit specialists are to be used to help develop and/or review plans for sampling applications. As more persons within the practice unit become experienced in using statistical sampling, the use of specialists can be restricted to those audit problems that require sophisticated statistical procedures.

How to Use Statistical Sampling

Auditors in the field need guidelines for determining how to use statistical sampling in their audit engagement. The practice unit must address the important question of what form these guidelines should take. Rigid guidelines that specify in detail what must be done may have the undesirable consequence of making the application of statistical sampling a mechanical process. When this occurs, the auditor may feel relieved of any responsibility for thinking statistically and, therefore, lose many of the benefits of using statistical sampling.

On the other hand, it is not reasonable to expect busy auditors to engage in the rather complex planning analysis presented in chapter 7 and illustrated in chapter 8 without some help. This suggests that the policy statement should set guidelines not for the results but for the *process*. Forms may be devised to aid the auditors in the field to implement the process described in the policy statement.

The policy statement might describe the planning process, including how to determine tolerable levels of sampling risk for both compliance and substantive tests and how to determine an amount to be considered material. Additionally, the policy statement might describe acceptable methods for evaluating statistical evidence. This will entail selecting either a positive or negative approach as described in chapter 3 and, perhaps, specifying the requirements for any adjustments to the recorded amounts based on a statistical sample.

Training

The first general standard of GAAP requires adequate technical training and proficiency for conducting any audit procedure. To implement this standard with respect to statistical sampling, the practice unit needs to decide what constitutes adequate training and proficiency in statistical sampling and how to achieve that prescribed level. Conforming to the first general standard does not require each person within the practice unit to become an expert in statistical sampling, but there is some minimum proficiency level that should be attained by anyone using statistical sampling. Moreover, within the practice unit, some persons should have a level of proficiency well above the minimum.[1]

Objectives

The minimum proficiency level was stated previously in terms of the following four areas: (1) the benefits and limitations of the alternative sampling methods permitted by the practice unit, (2) the proper application of the more commonly used methods, (3) the relation of sample results to audit conclusions, and (4) the documentation of applications. The following description elaborates somewhat on the first three of these objectives—the fourth is discussed separately later in the chapter.

Benefits and Limitations. As a minimum, the audit team needs to understand the benefits as well as the limitations of the alternative sampling methods that may be employed in an audit situation. The practice unit may limit the available alternatives that need to be considered through its policy statement. Some practice units may prescribe a single sampling method that is to be used whenever it is appropriate. In such a case, persons need to know what set of circumstances could lead to inappropriate usage.

When the audit team has a choice of methods, the policy statement may prescribe some guidelines for selecting the method to use. In such a circumstance, the audit team needs to know how to apply the guidelines in specific situations.

For example, in the case of compliance tests, the principal statistical method is some form of attribute sampling. The possible alternatives are between using unrestricted random sampling or a pps sample. In addition, there may be a choice between using a fixed sample size or some form of sequential sampling.

---

1. Many practice units rely on outside experts to help resolve the more technical problems. This helps to define the level of proficiency required within the practice unit.

For substantive tests, the potential range of sampling methods is rather broad. The two most widely used selection methods are a stratified random sample based on recorded amounts and a pps sample. With either method of selection, there are choices of evaluation technique. Assuming that the policy statement broadly describes the situation where one or more methods are valid, the audit team requires training in recognizing those situations in practice.

Proper Application. The audit team should be able to apply those statistical methods specified in the policy statement. Regardless of the specific statistical method used, proper application entails defining the objectives, population, frame, and sampling unit. For statistical methods used to test compliance, the attributes, threshold rates of unsatisfactory compliance, and risks of unwarranted reliance need to be specified. In addition, the risk of overauditing can also be specified, if desired. For substantive tests, the amount considered material, the tolerable sampling risk, and the risk of overauditing need to be specified.

Applying statistical sampling will usually entail using computer routines both in planning and evaluation. The training program should provide experience in using these programs in accordance with the practice unit's policy statement.

Relation of Sample Results to Audit Conclusions. The chief objective of sampling the details of a balance or class of transactions is to provide evidence to support an audit conclusion. For this reason, an important training objective is to be able to use statistical evidence properly in reaching audit conclusions.

Appropriate planning plays a crucial role in being able to reach a useful audit conclusion. The audit team needs to understand the necessity for formulating their audit objectives in operational terms and selecting the corresponding statistical objectives. For example, in testing sales, the operational objective of "Examination of independent shipping documents to determine whether all goods shipped have been invoiced" is far better than the broadly stated objective of "Test sales to see that they are properly recorded."

In addition to planning, the audit team should be able to evaluate their results statistically and relate those results to their audit conclusions. This is necessary both when the statistical results correspond to what was anticipated and when the results contain surprises.

Moreover, the qualitative aspects of the statistical evidence require emphasis. Determining reasons for the occurrence of errors—either compliance errors or monetary errors—is of utmost importance whether or not samples are selected statistically. The experienced members of the audit team need reassurance that statistical sampling does not lessen the need for these types of analysis, while the inexperienced need training in the necessity for doing such qualitative analysis.

Methods

How can these training objectives be achieved? A proper answer to this question must consider both the short-run and the long-run. In the short-run, the problem concerns ways of bringing all audit team members up to a minimum standard of competence and making sure that the practice unit has enough technical competence to achieve an acceptable quality standard in its applications of statistical sampling. In the long-run, the problem concerns the appropriate division of responsibility between the colleges and universities on the one hand and the profession on the other.

Although most recent college graduates have completed a basic statistics course as part of their degree requirements, few have received the grounding necessary in sampling theory to understand adequately the application of statistical sampling to audit problems. This means that entrants into the profession need training that emphasizes the application of statistical sampling to auditing problems.

Experienced auditors who have been out of school for several years probably remember little of any statistics they might have studied at one time. Some may never have had any formal training in statistical methodology. Training this group of persons may require some introduction to the fundamentals of statistical sampling before applications can be meaningfully discussed. The self-study program of the AICPA is one example of available material that can be used to introduce some of the fundamental concepts to those with little or no background in statistics.

Short-run Programs. In the short-run, a practice unit will need to provide some training for each person on the audit team. Some practice units have designed their own formal training programs for this purpose. These programs vary in length from three to five days. This variability is partially explained by two factors: the leeway each audit team has in making decisions relative to the sampling plan and/or whether there are individuals within the practice unit who function as statistical auditing specialists.

A typical training program provides for some prior self-study and a combination of lecture/discussion and case studies. The self-study portion of the course may require up to 40 hours of work including such material as the AICPA volumes, section 320 of SAS no. 1 and other selected reading from the professional literature or internal publications.

The lecture/discussions and case studies can be designed to allow participants to make the types of decisions they will be expected to make on their audit engagements. Case studies that closely simulate actual situations are most beneficial in accomplishing this. Cases can be divided into those that emphasize planning, execution, and evaluation. Planning cases can range from planning a particular application, such as confirmation of accounts receivable, to planning the entire audit engagement. The case discussed in chapter 8 is somewhere between these extremes—it encompasses the sales/receivables cycle.

Cases emphasizing planning should require the participants to decide the appropriate risk levels, amount of materiality, and appropriate sampling methods. In addition, cases should require the participants to define objectives, the population, frame, and sampling unit for individual tests. Cases covering compliance applications should emphasize the need for properly defining an occurrence and specifying appropriate threshold rates for unsatisfactory compliance.

Those cases emphasizing execution should reinforce the necessity for carefully following the sampling plan to reduce the possibility that nonsampling errors may occur. In addition, these cases offer a good opportunity to stress the need for carefully prepared working papers.

Cases emphasizing evaluation should cover both situations where the results correspond to what was anticipated in the planning phase as well as what to do when surprises occur. One pervasive problem that requires attention is what to do when a document is missing or when a response to a confirmation request cannot be obtained. Another concerns the need for thorough error analysis for each type of error found in a sample. Participants should gain experience in considering alternative sources of evidence when the sample does not corroborate the recorded amounts, including the statistical evidence that may be required to support an adjustment to the records.

Very little, if any, of the time needs to be spent making the computations. It is far more important that results be properly interpreted than that a person be able to compute a standard error. Moreover, any practice unit that decides to use statistical sampling will find it beneficial, and probably necessary, to have a battery of computer programs, such as are described in chapter 9. Some practice in using such programs would normally be an integral part of the training program.

Discussion leaders for the type of program described above should probably be experienced auditors who have extensively used statistical sampling. Some practice units have created effective discussion teams by combining an experienced auditor who has used statistical sampling with a person who has been extensively trained in statistics and has had experience in applying statistical sampling to auditing problems.

In addition to a formal training program, there should be opportunity for on-the-job training and supervision. On-the-job training is necessary as reinforcement to the formal classroom training. The practice unit must decide how this can best be accomplished. Some may select a formal procedure, others may adopt an informal system. In either case, the objectives of the training program cannot be met without some practical experience using statistical sampling.

Some practice units have found it advantageous to designate certain persons as statistical auditing specialists. Such persons generally have had a good academic background in statistics and extensive experience in applying statistical techniques to auditing problems. Since the function of these specialists is to assist in solving unusual applications as well as reviewing both plans and evaluations, their training should afford an opportunity to develop these skills.

Long-run Programs. In the long-run, the colleges and universities need to accept more responsibility for educating their students in the fundamentals of statistical sampling as it pertains to auditing. The basic statistics course required in many academic programs needs to be reinforced with material pertaining more directly to sampling in the accounting environment. The emphasis in this new material should be on applying statistical principles to obtain evidence based upon samples.

With such training, practice units could then concentrate their training efforts on their particular policies and procedures. A training program emphasizing cases presented to persons with a couple of years of auditing experience would continue to be beneficial, but the level of sophistication could be somewhat higher than is possible in today's environment.

As statistical procedures become more of an integral part of auditing practice, the need for separate statistical training programs will diminish. The practice unit can incorporate necessary statistics training within the framework of its regular program of professional development.

Documentation

The need for appropriate documentation is a professional requirement. Documentation of a sampling application requires working papers that reflect the three phases of planning, execution, and evaluation. Some practice units have found it advantageous to design forms for recording the required information pertaining to each sampling application. Whether or not forms are used, the documentation should be sufficient to permit a reviewer to determine whether the application represents a valid use of statistics and whether the audit conclusions are justified.

For planning purposes, documentation should include the following:

Audit objectives
Description of sampled population
Description of frame and sampling unit
Tolerable risks (both sampling risk and the risk of overauditing)
Minimum material amount (for substantive tests) or threshold rate for unsatisfactory compliance (for compliance tests)
Desired reliability (one-sided or two-sided)
Desired precision (upper precision limit for attributes test)
Sample size and how determined
Sample selection method
Anticipated evaluation technique

Listing these items is necessary but not sufficient for satisfactory documentation. The working papers should also contain a brief description of the criteria used for such items as the definition of the population, frame, and sampling unit, the tolerable risks, the minimum material amount, and the choice of sampling method.

The documentation of the execution phase should include a description of what was done as well as how any problems were resolved, such as failure to locate a particular document.

The documentation of the evaluation phase should include both the statistical results of the test as well as the audit conclusions reached. The statistical results for variables may be reported using either the positive or the negative approach. In either case, the working papers should indicate the achieved precision as well as any adjustment that was made to maintain the tolerable sampling risk at the planned level.

Audit conclusions should be explicitly stated. When the statistical evidence of a compliance test indicates that compliance is satisfactory, the auditor concludes that the statistical evidence corroborates his planned degree of reliance. On the other hand, when the statistical evidence fails to indicate satisfactory compliance, the working papers should state what changes were made in the preliminary audit program. Failure to document the changes may suggest that the statistical evidence was ignored.

A similar requirement exists for a substantive test. When the statistical evidence supports the recorded amount, the auditor may conclude that the recorded amount is fairly stated with respect to the objectives of the particular test. However, when the results fail to support the reasonableness of the recorded amount, the working papers should indicate what additional audit procedures were used to support the audit conclusion.

Review Process

A statistical application should be reviewed both prior to execution and after evaluation. The purpose of the first review is to ascertain whether the plan is statistically valid and cost effective and, if carried out, will lead to useful audit conclusions. The subsequent review should ascertain whether the plan was properly executed and the results appropriately evaluated.

To be effective such preliminary approval and subsequent review should be performed by persons experienced both in auditing and statistical sampling. For many applications this requirement can be met by audit executives who have received adequate training and have had experience in statistical applications. For some applications, the audit executive may need the assistance of a statistical audit specialist to ascertain whether the proposed plan is valid or whether the statistical evaluation was proper.

Many practice units that have adopted statistical sampling have found it desirable to designate a person as the coordinator of statistical sampling. This person is responsible for maintaining the quality of statistical applications, keeping the practice unit abreast of developments that will improve the usefulness of statistical techniques to the auditor and answering requests for assistance from those who need help in solving particular audit problems. In larger practice units, this function may be somewhat decentralized by designating persons within regions or offices to share some of these responsibilities.

The coordinator would also have responsibility for supervising the activities of any statistical audit specialist. The use of such specialists may be helpful in maintaining the quality of the statistical applications at a high level as well as providing needed assistance on more complex applications.

Common Problem Areas

In this section a discussion is presented of a few problems that are frequently encountered by practice units using statistical sampling.

Missing Documents

Sometimes during the course of a statistical test a document cannot be located; for example, in a test of disbursements, it may not be possible to find a particular purchase order. Normal auditing procedure would require the auditor to ascertain the reason for the document's absence. In the circumstances where investigation fails to reveal any specific reason, the auditor may first determine whether there is any acceptable alternative evidence. The criteria for acceptability is completely up to the auditor—the only statistical requirement is that an audited value be established for each sampling unit.

In the example of a missing purchase order, the auditor would consider whether there is any acceptable alternative source of evidence that would enable him to determine, say, whether the purchase had been properly authorized. If no such alternative source exists, the auditor must evaluate the missing evidence statistically.

Statistical evaluation involves both estimating the extent of missing documents and determining the possible effects that missing documents might have on the auditor's decision.

The most conservative evaluation is to regard any missing document as being completely in error. When doing this enables the auditor to conclude that compliance is satisfactory or that no material monetary error exists, there is nothing further required from a statistical viewpoint. When the conservative evaluation leads to the conclusion that compliance may not be satisfactory, but compliance would have been satisfactory if the missing document produced evidence of compliance, some further action is required.

One course of action for compliance tests would be to adjust the audit program to reflect less than the planned reliance on the particular control procedures for which the evidence is missing. This is the safest alternative. A somewhat less satisfactory alternative would be to expand the size of the compliance test. This alternative is not completely satisfactory, however, because it involves a form of sequential sampling that increases the sampling risk somewhat above the nominal level.

If the auditor decides to expand the sample, he should select enough additional items to maintain the sampling risk close to the planned level. One way of accomplishing this is to use the number of observed occurrences, regarding the missing document as an occurrence, to determine a required total sample size. For example, suppose the auditor had taken a sample of 90 based upon an anticipated proportion of .02, a desired upper precision limit of .07, and a one-sided reliability of .95. If the sample contained two occurrences and one missing document, the auditor would determine the sample size required for an anticipated proportion of .03, a desired upper precision limit of .07, and a one-sided reliability of .95. From the tables, the required sample size is 160, and, consequently, the auditor would need 70 additional observations.

Expanding the test in this manner is generally acceptable even though the sampling risk may be somewhat above the nominal risk. In the above example the sampling risk is .0534 instead of .05.

Non-response

Occasionally the auditor will not be able to obtain a response to a confirmation request in spite of sending a second and maybe even a third request. The normal course of action is to examine internal evidence through "alternative procedures." When these alternative procedures allow the auditor to determine the correct or audited amount for the balance owed as of the confirmation date, there is no statistical problem. On the other hand, when the auditor's alternative procedures do not result in determining an audited amount, there is an evaluation problem.

One conservative evaluation procedure is to regard such an account balance as being 100 percent overstated. Such overstatement errors are then evaluated together with the observed monetary errors. When this evaluation results in the sample evidence supporting the recorded accounts receivable balance as being fairly stated with regard to existence and recorded amount, no further statistical evaluation is required.

This conservative evaluation procedure may lead to the statistical conclusion that the recorded balance could be materially misstated. In this circumstance, two alternatives are suggested:

1. Follow the procedure suggested by Loebbecke and Neter in their article, "Statistical Sampling in Confirming Receivables" (cited in the Selected Bibliography) to estimate the potential non-response in the population. Use this estimate as the basis for deciding whether additional audit procedures are warranted or whether the accounts receivable are auditable.
2. Increase the sample size sufficiently so that even when the unconfirmed balances are regarded as being 100 percent overstated, the upper precision limit of error is less than a material amount.

Auditors selecting the second alternative (increasing the sample size) should be aware that the resulting sampling risk may be somewhat higher than the nominal risk. This occurs because the auditor is using the results of his sample to determine the sample size and is similar to the increase discussed under missing documents.

Timing

When sample items are selected several days prior to the observation date, it is very likely that the recorded amounts will have changed between the two dates. For example, suppose that a sample of inventory items is selected on December 23 and the quantities of the sampled items are to be observed on December 31. If the auditor compares the observed quantities as of December 31 to the recorded quantities as of December 23, there may be differences caused by transactions that occurred between the two dates. What can be done to prevent such ordinary transactions from influencing a decision about the accuracy of the recorded amounts?

The answer is easiest when the details of any intervening transactions are available. In this case, the auditor can determine the quantity on hand as of the date of observation, adjust the quantity for transactions that occurred between December 23 and December 31, and compare this adjusted quantity with the recorded quantity as of December 23. Any difference between the two quantities will be regarded as pertaining to the inventory as of December 23 even though there is the possibility that the difference occurred after that date.

When the details of transactions between the two dates are not available, the auditor must resort to a less accurate method. The observed quantities extended by audited prices provide a basis for using a difference, ratio, or regression estimate of the *difference* between the inventory value as of December 31 and the recorded amount as of December 23. Comparing this estimate to the difference in the recorded amounts for the two dates allows the auditor to estimate the amount of error in the recorded amount caused by quantity errors. For example, if the recorded amount as of December 23 is $9 million, and it is $9,800,000 as of December 31, then the recorded net addition to the inventory is $800,000 ($9,800,000 − $9,000,000). Suppose the sample resulted in an estimated difference between the recorded as of December 23 and the actual as of December 31 of $600,000. Then the estimated difference attributed to recording errors is −$200,000 ($600,000 − $800,000).

Some of this difference is caused by sampling error, but because the transaction details are not available, the auditor can do no better. The standard error of the estimate is also larger than it would be if the transaction details were available.

What Is an Error?

An explicit definition of an error is necessary for any application of statistical sampling. This is so whether the error refers to a compliance deviation or to the difference between an audited and recorded amount. The selection of the appropriate definition should be done during the planning phase to correspond with the stated audit objectives.

For example, if the audit objective is to test an inventory to determine whether there could be material monetary error caused by using incorrect prices, the auditor needs to define an error as any difference between the recorded price and the audited price. As an alternative, the auditor might elect to test an inventory for both prices and quantities. In this case, an error would be any difference in price or quantity.

Similarly, if the audit objective of a test of accounts receivable is to estimate the net realizable value, the auditor's definition of an error needs to include all possible sources of difference, such as an account that does not exist, a recorded account balance that is incorrect, or a recorded account balance that will not be collected.

Defining an error in a compliance test requires the auditor to make sure that the definition used will permit a useful audit conclusion. In addition, the acceptable evidence of compliance needs to be carefully stated. If the error is lack of proper authorization, the auditor needs to define operationally exactly what constitutes evidence of proper authorization. If one type of acceptable evidence is an approving signature on a purchase order, the lack of an authorized signature would be an occurrence. (See chapter 7, *Defining the attribute,* p. 149, for additional discussion.)

Appendixes

Appendix 1

Glossary of Terms

acceptance sampling A statistical procedure for deciding whether a manufactured lot of items is acceptable based on examining a sample from the lot. A related technique useful in auditing is called *attribute decision* in this book.

achieved precision The precision amount computed from the sample results. It will equal the planned precision only when the achieved standard error equals the standard error used in sample planning.

adjusted precision The precision amount calculated to have the same beta risk as expressed in the planned precision.

alpha risk As used in this book, the risk of deciding that there could be a material amount of monetary error when, in fact, the recorded amount is correct.

attribute A qualitative characteristic, such as absence of an authorized signature, that is associated with a sampling unit.

attribute decision A statistical procedure for deciding whether the rate of compliance deviation is as large as a stipulated threshold rate.

audited amount The amount established by the auditor as representing the amount that should be in the client's records relative to the particular sources of misstatement tested by his audit procedures.

audit risk The risk that material errors or irregularities, if they exist, will not be detected.

beta risk As used in this book, the risk of deciding that the recorded amount is correct when, in fact, the recorded amount is materially misstated.

bias Sample selection is *biased* whenever any sampling unit has a probability of being selected that is different from the planned probability.

binomial distribution The approximate sampling distribution for the sample occurrence rate (p) when an unrestricted random sample of size n is selected without replacement from a population of size N. If $P = M/N$ denotes the unknown population occurrence rate, and $p = m/n$, then

$$Prob\{p = a\} = \binom{n}{an} P^{an}(1 - P)^{n-an}$$

$$a = 0, \frac{1}{n}, \frac{2}{n}, \ldots, 1.$$

The approximation is appropriate whenever the sampling fraction (n/N) is less than .1.

characteristic A number or quality associated with each sampling unit. A numerical characteristic is called a *variable.* A qualitative characteristic is called an *attribute.*

coefficient of variation The ratio of a standard deviation to the corresponding mean. The population coefficient of variation measures the relative dispersion of the population distribution.

correlation A measure of the degree to which two quantities are linearly related. The correlation coefficient (ρ) is a number between -1 and $+1$. The extremes (-1, $+1$) signify an exact linear relationship, while zero means that the two quantities have no linear relationship. The population correlation coefficient may be computed from the following formula:

$$\rho = \frac{\sum_{1}^{N} (X_j - \bar{X})(Y_j - \bar{Y})}{\sqrt{\sum (X_j - \bar{X})^2 \sum (Y_j - \bar{Y})^2}}.$$

The sample estimate of the correlation is

$$\rho = \frac{\sum_{1}^{n} (x_j - \bar{x})(y_j - \bar{y})}{\sqrt{\sum (x_j - \bar{x})^2 \sum (y_j - \bar{y})^2}}.$$

decision interval The decision interval ranges from the recorded amount minus the precision to the recorded amount plus the precision.

decision objective Using a decision objective, the auditor's primary objective is to decide whether the unknown population characteristic is within satisfactory limits.

degrees of freedom The index used to identify the particular sampling distribution for an estimate. The degrees of freedom for the mean estimator is one less than the sample size. For the difference and ratio estimates, the degrees of freedom may be set equal to one less than the number of observed differences. For the regression estimate, the degrees of freedom may be set equal to two less than the number of differences. For stratified random sampling, the number of degrees of freedom is reduced from the unstratified case by one fewer than the number of strata.

delta risk The symbol (δ) used to designate the sampling risk of unwarranted reliance in statistical compliance tests.

detection controls Those pertinent controls that have been designed to detect and correct any errors or irregularities that may have occurred as transactions are processed.

difference The population difference equals the total audited amount (X) minus the total recorded amount (Y). For an individual sampling unit, the difference equals the audited amount (x_j) minus the recorded amount (y_j).

discovery sampling (exploratory sampling) A procedure for determining the sample size required to have a stipulated probability of observing at least one occurrence when the population occurrence rate is at a designated level.

error analysis A determination of the cause of any observed compliance deviation or monetary difference.

estimation objective Using an estimation objective, the primary purpose of a statistical sample is to estimate an unknown population characteristic.

frame A listing of the sampling units.

hypergeometric distribution The exact sampling distribution for the sample occurrence rate (p) when an unrestricted random sample of size n is selected without replacement from a population of size N. If $P = M/N$ denotes the (unknown) population occurrence rate and $p = m/n$ denotes the sample occurrence rate, then

$$\text{Prob}\,\{p = a\} = \frac{\binom{PN}{an}\binom{(1-P)N}{(1-a)n}}{\binom{N}{n}}$$

$$a = 0, \frac{1}{n}, \frac{2}{n}, \ldots, 1$$

$$\text{where} \binom{N}{n} = \frac{N!}{(N-n)!n!}.$$

likelihood As used in this book, a subjective estimate of a probability of occurrence.

lower precision A measure of how much the unknown population characteristic may fall below the estimated amount at a specified one-sided reliability. Equals the standard error of the estimate multiplied by the corresponding one-sided reliability factor.

negative approach The auditor using the negative approach decides there may be material misstatement only when the achieved upper limit of monetary error exceeds a material amount—otherwise the auditor acts on the basis that the recorded amount is not materially in error.

nonsampling risk The portion of audit risk of not detecting a material error that exists because of the inherent limitations of the procedures used, the timing of the procedures, the system being examined, and the skill and care of the auditor.

normal distribution A symmetric bell-shaped frequency distribution that is the approximate sampling distribution for many statistical estimates. The normal distribution is completely determined by its mean and standard deviation. The standard normal distribution has mean 0 and standard deviation 1 and is tabled.

optimal allocation (Neyman allocation) A method for allocating a stratified sample to the strata. The stratum sample size is proportional to the product of the number of stratum population items times the stratum standard deviation. The stratum standard deviation used should correspond to the estimation technique employed. The formula is as follows:

$$n_i = n \frac{N_i \sigma_i}{\sum N_i \sigma_i},$$

where n is the total sample size, N_i is the number of population items within the ith stratum, and σ_i is the stratum standard deviation.

pertinent procedures "Pertinent procedures are those which, if not purported to be in use, would have affected adversely the auditor's preliminary evaluation of the system prior to his tests of compliance" (section 320B.15, SAS no. 1).

planned precision The precision amount used in sample planning.

Poisson distribution The Poisson distribution is a useful approximation for the sampling distribution of the number of occurrences in an unrestricted sample from a large population, when the occurrence rate is low. Tables of the Poisson distribution are indexed by a parameter (λ) so that $\lambda = nP$ where n is the sample size and P is the population occurrence rate.

population The aggregate of accounting entries about which information is desired.

population mean The center of the population distribution computed by adding the characteristics of each sampling unit and then dividing by the total number of sampling units. When the characteristic is an attribute, the population mean is the *population occurrence rate.*

population occurrence rate The proportion or percentage of the sampling units in the population that have the specified attribute.

population standard deviation A measure of the dispersion about the population mean.

positive approach The auditor using the positive approach decides the recorded amount is not materially in error only when the recorded amount is within the achieved precision interval—otherwise the auditor takes action to investigate potential material error.

pps Refers to a selection technique known as *probability proportional to size.* With this technique, when the measure of size is the recorded amount, each sampling unit has a probability of being selected that is proportional to its recorded amount. Three ways to select a pps sample are described in chapter 2, and the computer routine is described in chapter 9.

PRA allocation (proportional to recorded amount) A method for allocating a stratified sample to the strata. The stratum sample size equals the total sample size multiplied times the ratio of the total stratum recorded amount to the total recorded amount in all strata. The formula is as follows:

$$n_i = n \frac{Y_i}{Y},$$

where n is the total sample size, Y_i is the recorded amount of the ith stratum, and Y is the total recorded amount over all strata.

precision A measure of closeness between a sample estimate and the corresponding unknown population characteristic. It is computed by multiplying the standard error of the estimate by a factor (reliability factor) corresponding to the desired reliability. The reliability factor may be either one-sided or an interval. The one-sided factor leads to a one-sided precision and the interval reliability factor leads to a precision interval.

precision interval Formed by adding and subtracting the precision at a specified interval reliability to the sample estimate.

prevention controls Those pertinent controls that have been designed to prevent errors or irregularities from occurring.

probability Refers either to the relative frequency of occurrence of some event or to a person's subjective measure of belief that some event will occur.

ratio The population ratio equals the total audited amount (X) divided by the total recorded amount (Y).

The sample ratio equals the estimated total audited amount ($\hat{X}$) divided by the estimated total recorded amount ($\hat{Y}$).

recorded amount The amount appearing in the client's records.

relative precision The ratio of the precision to the amount being estimated.

reliability (confidence), one-sided and interval The *interval* reliability measures the proportion of all such precision intervals that would contain the unknown population characteristic.

The *one-sided* reliability measures the proportion of upper precision limits that exceeds the population characteristic or the proportion of lower precision limits that fall below the population characteristic.

replicated sampling A sampling procedure that selects a sample of size n by selecting k subsamples, each of size m ($n = km$). Each subsample is statistically evaluated and the resulting k estimates are used to estimate the standard error of the pooled estimate based on all n observations. For details, see Deming [7] and Cochran [5].

risk of overauditing For substantive tests, this is the risk that the statistical test indicates that there may be a material error when, in fact, there is none.

For compliance tests, this is the risk that the statistical test indicates that compliance may be unsatisfactory when, in fact, it is satisfactory.

risk of unwarranted reliance The risk that the auditor relies on the pertinent controls to a greater extent than he would if he had complete knowledge of the effectiveness of the procedures.

In testing compliance statistically, the risk of unwarranted reliance is the risk that the auditor decides compliance is satisfactory when, in fact, the rate of compliance deviations equals the threshold rate for unsatisfactory compliance.

sample mean Equals the sum of the sample values divided by the sample size. The sample mean is an unbiased estimate of the population mean.

sample standard deviation A measure of the dispersion of the sample and an estimate of the population standard deviation.

sampling distribution Describes the probability of each possible value of the sample point estimate. The sampling distribution depends upon (1) the sample size, (2) the method of selection, (3) the observed characteristic, (4) the evaluation procedure, and (5) the population distribution.

sampling risk The portion of audit risk of not detecting a material error that exists because the auditor examined a sample of the account balances or transactions instead of every one.

sampling unit The individual members of the population.

sequential sampling A procedure for reaching statistical decisions that does not specify the required sample size in advance. Sampling units are examined individually or in small groups until the cumulated statistical evidence is sufficient for a definite decision.

skewness A lack of symmetry about the population mean in a frequency distribution. Typical accounting populations are skewed because there are many small to medium amounts and several very large amounts. The skewness of a population is measured by the following factor:

$$G_1 = \frac{1}{N\sigma_Y{}^3} \sum_{1}^{N} (Y_j - \bar{Y})^3.$$

standard error of the estimate The standard deviation of the sampling distribution corresponding to a particular estimating procedure is called the standard error of the estimate.

statistical efficiency One statistical technique is said to be more efficient than another if it requires a smaller sample size to achieve the same precision and reliability.

Student's *t*-distribution The approximate sampling distribution for the ratio of the sample mean to the sample standard deviation when the sample size is not large. It is the exact sampling distribution only when the population distribution is normal. The distribution is indexed by the number of degrees of freedom.

systematic sampling A method of selecting a sample of size n from a population of size N by first selecting a random number between 1 and $[N/n]$ and then selecting every $[N/n]$th unit from the entire frame. $[N/n]$ refers to the smallest integer less than or equal to N/n. If k random starts are desired, then k numbers are selected between 1 and $[kN/n]$ and the skip interval is $[kN/n]$.

threshold rate for unsatisfactory compliance Represents the lowest rate of compliance deviation that would cause the auditor to use less than his planned degree of reliance.

two-stage sampling A sampling technique in which a sample of large units (primary units) is selected at the first stage. At the second stage a sample of elements is selected from each of the selected primary units. This technique is also called *subsampling*. See Cochran [5], chapters 10 and 11.

unbiased A statistical estimator is said to be unbiased if its average value taken over all possible samples equals the corresponding population amount.

unrestricted random sampling A method of selection in which each sampling unit has an equal probability of being selected and each group of n units has an equal probability of being selected. The term n refers to the sample size.

upper precision A measure of how much the unknown population characteristic may exceed the estimated amount at a specified one-sided reliability. Equals the standard error of the estimate multiplied by the corresponding one-sided reliability factor.

variable A quantitative characteristic, such as a dollar amount, that is associated with a sampling unit.

Appendix 2

List of Formulas

Unstratified Mean Estimation

Formula	*Description*
$\bar{x} = \frac{\sum x_j}{n}$	Sample mean of audited amounts
$\hat{X}_M = N\bar{x}$	Unstratified mean estimator of total audited amount
$S_X = \sqrt{\frac{\sum x_j^2 - n\bar{x}^2}{n-1}}$	Estimated standard deviation of audited amounts
$\hat{\sigma}(\hat{X}_M) = \frac{NS_X \sqrt{1 - n/N}}{\sqrt{n}}$	Estimated standard error of unstratified mean estimator
$n = \frac{N^2U_R{}^2S_X{}^2}{A^2 + NU_R{}^2S_X{}^2}$	Sample size formula for unstratified mean estimator
$A'_M = \frac{NU_RS_X \sqrt{1 - n/N}}{\sqrt{n}}$	Achieved precision of unstratified mean estimator

Unstratified Difference Estimation

Formula	*Description*
$\hat{D} = N\bar{d}$	Unstratified difference estimator of total difference
$\hat{X}_D = Y + \hat{D}$	Unstratified difference estimator of total audited amount
$S_D = \sqrt{\frac{\sum d_j^2 - n\bar{d}^2}{n-1}}$	Estimated standard deviation of difference amounts
$\hat{\sigma}(\hat{D}) = \frac{NS_D \sqrt{1 - n/N}}{\sqrt{n}}$	Estimated standard error of unstratified difference estimator

$n = \dfrac{N^2 U_R{}^2 S_D{}^2}{A^2 + N U_R{}^2 S_D{}^2}$	Sample size formula for unstratified difference estimator
$\sqrt{\hat{p}_D \sigma_Y{}^2 + \hat{p}_D(1 - \hat{p}_D)\overline{Y}^2}$	Approximate expected value of standard deviation of differences when all differences are 100 percent overstatements
$A'_D = \dfrac{N U_R S_D \sqrt{1 - n/N}}{\sqrt{n}}$	Achieved precision of unstratified difference estimator
$\hat{\sigma}_D = \sqrt{P_U(m) S_D{}^2(m) + P_U(m)(1 - P_U(m))\bar{d}_m{}^2}$	Estimated standard deviation of difference amounts (approximation)

Unstratified Ratio Estimation

Formula	*Description*
$R = \dfrac{\sum x_j}{\sum y_j} = 1 + \dfrac{\sum d_j}{\sum y_j}$	Estimated ratio of audited amount to recorded amount
$\hat{X}_R = \hat{R}Y$	Unstratified ratio estimator of total audited amount
$S_R = \sqrt{\dfrac{\sum x_j{}^2 + \hat{R}^2 \sum y_j{}^2 - 2\hat{R} \sum x_j y_j}{n - 1}}$ $= \sqrt{\dfrac{\sum d_j{}^2 + (\hat{R} - 1)^2 \sum y_j{}^2 - 2(\hat{R} - 1) \sum d_j y_j}{n - 1}}$	Estimated standard deviation of ratios
$\hat{\sigma}(\hat{X}_R) = \dfrac{N S_R \sqrt{1 - n/N}}{\sqrt{n}}$	Estimated standard error of unstratified ratio estimator
$n = \dfrac{N^2 U_R{}^2 S_R{}^2}{A^2 + N U_R{}^2 S_R{}^2}$	Sample size formula for unstratified ratio estimator
$\sqrt{\hat{p}_D(1 - \hat{p}_D)(\sigma_Y{}^2 + \overline{Y}^2)}$	Approximate expected value of standard deviation of ratios when all differences are 100 percent overstatements
$A'_R = \dfrac{N U_R S_R \sqrt{1 - n/N}}{\sqrt{n}}$	Achieved precision of unstratified ratio estimation
$\hat{\sigma}_R = \sqrt{P_U(m) S_R{}^2(m)}$	Estimated standard deviation of ratios (approximation)

Unstratified Regression Estimation

Formula	*Description*
$b = \dfrac{\sum x_i y_i - n\bar{x}\bar{y}}{\sum y_i^2 - n\bar{y}^2} = 1 + \dfrac{\sum d_i y_i - n\bar{d}\bar{y}}{\sum y_i^2 - n\bar{y}^2}$	Estimated regression coefficient
$\hat{X}_G = N\bar{x} + b(Y - N\bar{y})$	Unstratified regression estimator of total audited amount
$\hat{D}_G = N\bar{d} + (b - 1)(Y - N\bar{y})$	Unstratified regression estimator of total differences
$S_G = \sqrt{\dfrac{1}{n-2}\left[\sum x_i^2 - n\bar{x}^2 - \dfrac{(\sum x_i y_i - n\bar{x}\bar{y})^2}{\sum y_i^2 - n\bar{y}^2}\right]} = \sqrt{\dfrac{1}{n-2}\left[\sum d_i^2 - n\bar{d}^2 - \dfrac{(\sum d_i y_i - n\bar{d}\bar{y})^2}{\sum y_i^2 - n\bar{y}^2}\right]}$	Estimated standard deviation of regression amounts
$\hat{\sigma}(\hat{X}_G) = \dfrac{NS_G\sqrt{1 - n/N}}{\sqrt{n}}$	Estimated standard error of unstratified regression estimator
$n = \dfrac{N^2 U_R^2 S_G^2}{A^2 + N U_R^2 S_G^2}$	Sample size formula for unstratified regression estimator
$A'_G = \dfrac{N U_R S_G \sqrt{1 - n/N}}{\sqrt{n}}$	Achieved precision of unstratified regression estimator
$\hat{\sigma}_G = \sqrt{P_U(m)S_D^2(m) + P_U(m)(1 - P_U(m))\bar{d}_m^2 - \left(\dfrac{n-1}{n-2}\right)(b-1)^2 S_y^2}$	Estimated standard deviation of regression amounts (approximation)

Stratified Mean Estimation

Formula	*Description*
$\hat{X}_{MS} = \sum N_i \bar{x}_i$	Stratified mean estimator of total audited amount
$\hat{\sigma}(\hat{X}_{MS}) = \sqrt{\sum N_i(N_i - n_i)\dfrac{S_{xi}^2}{n_i}}$	Estimated standard error of the stratified mean estimator
$n = \dfrac{U_R^2(\sum N_i \sigma_{Yi})^2}{A^2 + U_R^2 \sum N_i \sigma_{Yi}^2}$	Sample size formula for the stratified mean estimator using optimal allocation

$$n = \frac{U_R^2 Y \sum N_i^2 \frac{\sigma_{Yi}^2}{Y_i}}{A^2 + U_R^2 \sum N_i \sigma_{Yi}^2}$$

Sample size formula for the stratified mean estimator using PRA allocation (proportional to the recorded amount)

$$A'_{MS} = U_R \sqrt{\sum N_i (N_i - n_i) \frac{S_{Xi}^2}{n_i}}$$

Achieved precision of the stratified mean estimator

Stratified Difference Estimation

Formula	*Description*
$\hat{D}_S = \sum N_i \bar{d}_i$	Stratified difference estimator of total difference
$\hat{X}_{DS} = Y + \hat{D}_S$	Stratified difference estimator of total audited amount
$\hat{\sigma}(\hat{D}_S) = \sqrt{\sum N_i (N_i - n_i) \frac{S_{Di}^2}{n}}$	Estimated standard error of stratified difference estimator
$n = \frac{U_R^2 (\sum N_i S_{Di})^2}{A^2 + U_R \sum N_i S_{Di}^2}$	Sample size formula for the stratified difference estimator using optimal allocation
$n = \frac{U_R^2 Y \sum N_i^2 \frac{S_{Di}^2}{Y_i}}{A^2 + U_R^2 \sum N_i S_{Di}^2}$	Sample size formula for the stratified difference estimator using PRA allocation
$A'_{DS} = U_R \sqrt{\sum N_i (N_i - n_i) \frac{S_{Di}^2}{n_i}}$	Achieved precision of the stratified difference estimator
$\hat{\sigma}_D(i) = \sqrt{P_U(m) S_D^2(m_i) + P_U(m)(1 - P_U(m)) \bar{d}_{m_i}^2}$	Estimated standard deviation of difference amounts in ith stratum (approximation)

Combined Ratio Estimation

Formula	*Description*
$\hat{R}_C = \frac{\sum N_i \bar{x}_i}{\sum N_i \bar{y}_i} = 1.0 + \frac{\sum N_i \bar{d}_i}{\sum N_i \bar{y}_i}$	Combined ratio estimator of population ratio of audited to recorded amount
$\hat{X}_{RC} = \hat{R}_C Y$	Combined ratio estimator of total audited amount

$$\hat{\sigma}(\hat{X}_{RC}) = \sqrt{\sum N_i(N_i - n_i)\frac{S_{RCi}^2}{n_i}}$$

Estimated standard error of the combined ratio estimator

$$S_{RCi} = \sqrt{\frac{\sum x_{ij}^2 + \hat{R}_C^2 \sum y_{ij}^2 - 2\hat{R}_C \sum x_{ij}y_{ij}}{n_i - 1}}$$

$$= \sqrt{\frac{\sum d_{ij}^2 + (\hat{R}_C - 1)^2 \sum y_{ij}^2 - 2(\hat{R}_C - 1)\sum d_{ij}y_{ij}}{n_i - 1}}$$

Estimated standard deviation of ratio in ith stratum

$$n = \frac{U_R^2(\sum N_i S_{RCi})^2}{A^2 + U_R^2 \sum N_i S_{RCi}^2}$$

Sample size formula for the combined ratio estimator using optimal allocation

$$n = \frac{U_R^2 Y \sum \frac{N_i^2 S_{RCi}^2}{Y_i}}{A^2 + U_R^2 \sum N_i S_{RCi}^2}$$

Sample size formula for the combined ratio estimator using PRA allocation

$$A'_{RC} = U_R\sqrt{\sum N_i(N_i - n_i)\frac{S_{RCi}^2}{n_i}}$$

Achieved precision of the combined ratio estimator

Combined Regression Estimation

Formula	*Description*

$$b_c = \frac{\sum N_i(N_i - n_i)\frac{S_{XYi}}{n_i}}{\sum N_i(N_i - n_i)\frac{S_{Yi}^2}{n_i}}$$

$$= 1 + \frac{\sum N_i(N_i - n_i)\frac{S_{DYi}}{n_i}}{\sum N_i(N_i - n_i)\frac{S_{Yi}^2}{n_i}}$$

Combined regression estimator of the regression coefficient

$$\hat{X}_{GC} = \sum N_i\bar{x}_i + b_c(Y - \sum N_i\bar{y}_i)$$

Combined regression estimator of total audited amount

$$S_{XYi} = \frac{\sum x_{ij}y_{ij} - n_i\bar{x}_i\bar{y}_i}{n_i - 1}$$

Estimated covariance between audited and recorded amounts in ith stratum

$$S_{DYi} = \frac{\sum d_{ij}y_{ij} - n_i\bar{d}_i\bar{y}_i}{n_i - 1}$$

Estimated covariance between difference and recorded amounts in ith stratum

$$\hat{\sigma}(\hat{X}_{GC}) = \sqrt{\sum N_i(N_i - n_i)\frac{S_{GCi}^2}{n_i}}$$

Estimated standard error of the combined regression estimator

$$S_{GCi} = \sqrt{S_{Xi}^2 - 2b_cS_{XYi} + b_c^2S_{Yi}^2}$$

$$= \sqrt{S_{Di}^2 - 2(b_c - 1)S_{DYi} + (b_c - 1)^2S_{Yi}^2}$$

Estimated standard deviation of regression amounts within ith stratum

$$n = \frac{U_R^2(\sum N_i S_{GCi})^2}{A^2 + U_R^2 \sum N_i S_{GCi}^2}$$

Sample size formula for the combined regression estimator using optimal allocation

$$n = \frac{U_R^2 Y \sum \frac{N_i^2 S_{GCi}^2}{Y_i}}{A^2 + U_R^2 \sum N_i S_{GCi}^2}$$

Sample size formula for the combined regression estimator using PRA allocation

$$A'_{GC} = U_R \sqrt{\sum N_i(N_i - n_i)\frac{S_{GCi}^2}{n_i}}$$

Achieved precision of combined regression estimator

$$\hat{\sigma}_{GC}(i) = \sqrt{P_U(m)S_{Di}^2(m_i) + P_U(m)(1 - P_U(m))\bar{d}_{m_i}^2 - 2(b_c - 1)S_{DYi} + (b_c - 1)^2 S_{Yi}^2}$$

Estimated standard deviation of regression amounts in *i*th stratum (approximation)

Appendix 3A

Normal Curve Areas, *U* Values, and Reliability Percentages

Normal Curve Areas

σ	.00	.01	.02	.03	.04	.05	.06	.07	.08	.09
0.0	.0000	.0040	.0080	.0120	.0159	.0199	.0239	.0279	.0319	.0359
0.1	.0398	.0438	.0478	.0517	.0557	.0596	.0636	.0675	.0714	.0753
0.2	.0793	.0832	.0871	.0910	.0948	.0987	.1026	.1064	.1103	.1141
0.3	.1179	.1217	.1255	.1293	.1331	.1368	.1406	.1443	.1480	.1517
0.4	.1554	.1591	.1628	.1664	.1700	.1736	.1772	.1808	.1844	.1879
0.5	.1915	.1950	.1985	.2019	.2054	.2088	.2123	.2157	.2190	.2224
0.6	.2257	.2291	.2324	.2357	.2389	.2422	.2454	.2486	.2518	.2549
0.7	.2580	.2612	.2642	.2673	.2704	.2734	.2764	.2794	.2823	.2852
0.8	.2881	.2910	.2939	.2967	.2995	.3023	.3051	.3078	.3106	.3133
0.9	.3159	.3186	.3212	.3238	.3264	.3289	.3315	.3340	.3365	.3389
1.0	.3413	.3438	.3461	.3485	.3508	.3531	.3554	.3577	.3599	.3621
1.1	.3643	.3665	.3686	.3708	.3729	.3749	.3770	.3790	.3810	.3830
1.2	.3849	.3869	.3888	.3907	.3925	.3944	.3962	.3980	.3997	.4015
1.3	.4032	.4049	.4066	.4083	.4099	.4115	.4131	.4147	.4162	.4177
1.4	.4192	.4207	.4222	.4236	.4251	.4265	.4279	.4292	.4306	.4319
1.5	.4332	.4345	.4357	.4370	.4382	.4394	.4406	.4418	.4430	.4441
1.6	.4452	.4463	.4474	.4485	.4495	.4505	.4515	.4525	.4535	.4545
1.7	.4554	.4564	.4573	.4582	.4591	.4599	.4608	.4616	.4625	.4633
1.8	.4641	.4649	.4656	.4664	.4671	.4678	.4686	.4693	.4699	.4706
1.9	.4713	.4719	.4726	.4732	.4738	.4744	.4750	.4758	.4762	.4767
2.0	.4773	.4778	.4783	.4788	.4793	.4798	.4803	.4808	.4812	.4817
2.1	.4821	.4826	.4830	.4834	.4838	.4842	.4846	.4850	.4854	.4857
2.2	.4861	.4865	.4868	.4871	.4875	.4878	.4881	.4884	.4887	.4890
2.3	.4893	.4896	.4898	.4901	.4904	.4906	.4909	.4911	.4913	.4916
2.4	.4918	.4920	.4922	.4925	.4927	.4929	.4931	.4932	.4934	.4936
2.5	.4938	.4940	.4941	.4943	.4945	.4946	.4948	.4949	.4951	.4952
2.6	.4953	.4955	.4956	.4957	.4959	.4960	.4961	.4962	.4963	.4964
2.7	.4965	.4966	.4967	.4968	.4969	.4970	.4971	.4972	.4973	.4974
2.8	.4974	.4975	.4976	.4977	.4977	.4978	.4979	.4980	.4980	.4981
2.9	.4981	.4982	.4983	.4984	.4984	.4984	.4985	.4985	.4986	.4986
3.0	.4986	.4987	.4987	.4988	.4988	.4988	.4989	.4989	.4989	.4990
3.1	.4990	.4991	.4991	.4991	.4992	.4992	.4992	.4992	.4993	.4993

U Values and Reliability Percentages

This chart of *U* values and reliability percentages can be used to convert either one from the other.

R	U_R	*R*	U_R
99.99%	3.87	83.84%	1.40
99.95	3.50	82.29	1.35
99.88	3.25	80.64	1.30
99.73	3.00	80.29	1.29
99.50	2.81	78.87	1.25
99.01	2.58	76.98	1.20
98.02	2.33	74.98	1.15
97.00	2.17	72.86	1.10
96.06	2.06	70.16	1.04
95.45	2.00	68.26	1.00
95.00	1.96	65.27	0.94
92.81	1.80	60.46	0.85
90.11	1.65	57.62	0.80
89.04	1.60	51.60	0.70
87.88	1.55	45.15	0.60
86.63	1.50	38.29	0.50
85.01	1.44	31.08	0.40

If trying to find a *U* value for a known *R* value which does not appear, use the *U* value for the next higher *R*. If trying to find *R* for a known *U* value which does not appear, use *R* for the next lower *U* value. Interpolating should not be done. See normal curve area table for more detailed values.

Appendix 3B

Student's *t*-Distribution

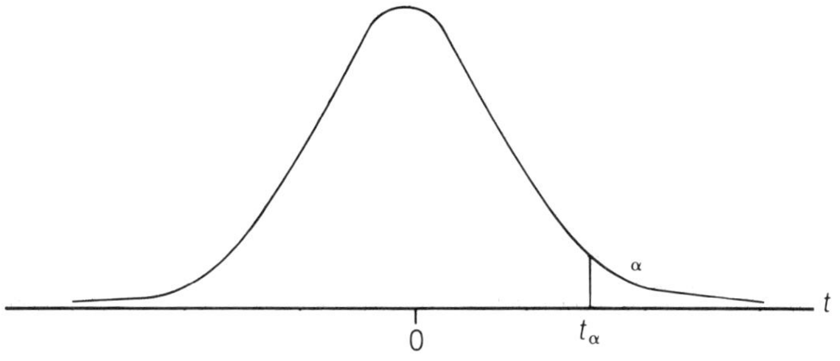

The following table provides the values of t_α that correspond to a given upper-tail area α and a specified number of degrees of freedom.

Degrees of Freedom	Upper-Tail Area α									
	.4	.25	.1	.05	.025	.01	.005	.0025	.001	.0005
1	0.325	1.000	3.078	6.314	12.706	31.821	63.657	127.32	318.31	636.62
2	.289	0.816	1.886	2.920	4.303	6.965	9.925	14.089	22.327	31.598
3	.277	.765	1.638	2.353	3.182	4.541	5.841	7.453	10.214	12.924
4	.271	.741	1.533	2.132	2.776	3.747	4.604	5.598	7.173	8.610
5	0.267	0.727	1.476	2.015	2.571	3.365	4.032	4.773	5.893	6.869
6	.265	.718	1.440	1.943	2.447	3.143	3.707	4.317	5.208	5.959
7	.263	.711	1.415	1.895	2.365	2.998	3.499	4.029	4.785	5.408
8	.262	.706	1.397	1.860	2.306	2.896	3.355	3.833	4.501	5.041
9	.261	.703	1.383	1.833	2.262	2.821	3.250	3.690	4.297	4.781
10	0.260	0.700	1.372	1.812	2.228	2.764	3.169	3.581	4.144	4.587
11	.260	.697	1.363	1.796	2.201	2.718	3.106	3.497	4.025	4.437
12	.259	.695	1.356	1.782	2.179	2.681	3.055	3.428	3.930	4.318
13	.259	.694	1.350	1.771	2.160	2.650	3.012	3.372	3.852	4.221
14	.258	.692	1.345	1.761	2.145	2.624	2.977	3.326	3.787	4.140
15	0.258	0.691	1.341	1.753	2.131	2.602	2.947	3.286	3.733	4.073
16	.258	.690	1.337	1.746	2.120	2.583	2.921	3.252	3.686	4.015
17	.257	.689	1.333	1.740	2.110	2.567	2.898	3.222	3.646	3.965
18	.257	.688	1.330	1.734	2.101	2.552	2.878	3.197	3.610	3.922
19	.257	.688	1.328	1.729	2.093	2.539	2.861	3.174	3.579	3.883
20	0.257	0.687	1.325	1.725	2.086	2.528	2.845	3.153	3.552	3.850
21	.257	.686	1.323	1.721	2.080	2.518	2.831	3.135	3.527	3.819
22	.256	.686	1.321	1.717	2.074	2.508	2.819	3.119	3.505	3.792
23	.256	.685	1.319	1.714	2.069	2.500	2.807	3.104	3.485	3.767
24	.256	.685	1.318	1.711	2.064	2.492	2.797	3.091	3.467	3.745
25	0.256	0.684	1.316	1.708	2.060	2.485	2.787	3.078	3.450	3.725
26	.256	.684	1.315	1.706	2.056	2.479	2.779	3.067	3.435	3.707
27	.256	.684	1.314	1.703	2.052	2.473	2.771	3.057	3.421	3.690
28	.256	.683	1.313	1.701	2.048	2.467	2.763	3.047	3.408	3.674
29	.256	.683	1.311	1.699	2.045	2.462	2.756	3.038	3.396	3.659
30	0.256	0.683	1.310	1.697	2.042	2.457	2.750	3.030	3.385	3.646
40	.255	.681	1.303	1.684	2.021	2.423	2.704	2.971	3.307	3.551
60	.254	.679	1.296	1.671	2.000	2.390	2.660	2.915	3.232	3.460
120	.254	.677	1.289	1.658	1.980	2.358	2.617	2.860	3.160	3.373
∞	.253	.674	1.282	1.645	1.960	2.326	2.576	2.807	3.090	3.291

SOURCE: E. S. Pearson and H. O. Hartley, *Biometrika Tables for Statisticians*, vol. 1 (London, 1966).

Appendix 3C

Sequential Sample Sizes

$P_1 = .005$

$\alpha \leq .05$

Risk of overauditing $\leq .05$

	P_0						
δ	.04	.05	.06	.07	.08	.09	.10
.01	128	99	81	68	59	51	46
.02	109	84	69	58	50	44	39
.03	98	76	61	52	45	39	35
.04	90	69	56	47	41	36	32
.05	83	65	53	44	38	33	30
.06	78	61	49	41	36	31	28
.07	74	57	47	39	34	30	26
.08	70	54	44	37	32	28	25
.09	67	52	42	35	31	27	24
.10	64	50	40	34	29	26	23
Δn	59	51	45	40	36	33	31

$P_1 = .01$

$\alpha \leq .10$

Risk of overauditing $\leq .05$

	P_0					
δ	.05	.06	.07	.08	.09	.10
.01			73	63	55	48
.02		75	62	53	46	41
.03		67	60	48	41	37
.04	78	62	51	44	38	34
.05	72	58	48	41	35	31
.06	68	54	45	38	33	29
.07	64	51	42	36	31	28
.08	61	49	40	34	30	26
.09	58	46	38	33	28	25
.10	56	44	37	31	27	24
Δn	40	36	32	29	27	25

Appendix 3D

Ratio of Precision to Materiality—Positive and Negative Approaches

Positive Approach

	Alpha Risk (1 – R)			
Beta Risk	.20	.10	.05	.01
.01	.355	.413	.457	.525
.025	.395	.456	.500	.568
.05	.437	.500	.543	.609
.075	.471	.532	.576	.641
.10	.500	.561	.605	.668
.15	.511	.612	.653	.712
.20	.603	.661	.700	.753
.25	.653	.708	.742	.791
.30	.707	.756	.787	.829
.35	.766	.808	.834	.868
.40	.831	.863	.883	.908
.45	.907	.926	.937	.952

Negative Approach

Beta Risk	Alpha Risk			
(1 – R_1)	.20	.10	.05	.01
.01	.645	.585	.543	.475
.025	.605	.543	.500	.432
.05	.563	.500	.457	.390
.075	.529	.466	.424	.358
.10	.500	.437	.395	.332
.15	.448	.387	.347	.287
.20	.396	.337	.300	.246
.25	.344	.289	.255	.206
.30	.289	.240	.210	.168
.35	.234	.191	.166	.131
.40	.163	.132	.113	.088
.45	.092	.073	.062	.048

Appendix 4

Technical Basis for Approximations

Suppose that all monetary differences are overstatements, each difference equals the recorded amount, and the differences are randomly distributed among the N population units. Under these assumptions, the N population units may be characterized as a realization of the following process:

$$D_j = \begin{cases} -Y_j & \text{with probability } p \\ 0 & \text{with probability } (1-p), \quad j = 1, \ldots, N. \end{cases}$$

This says that the particular set of differences in the population could have been generated by a Bernoulli process with the magnitude of the differences equal to the recorded amount.

Examination of the properties of this process leads to some useful conclusions concerning the set of realizations. These conclusions will be stated in terms of averages over all possible realizations. The averaging operation is denoted by the symbol E_p.

Proposition 1 $\quad E_p(\sigma_D^2) \doteq p[\sigma_Y^2 + (1-p)\bar{Y}^2]$.

This proposition says that the square of the standard deviation of the differences averaged over all realizations approximately equals the expected proportion (p) of differences times the square of the standard deviation of recorded amounts (σ_Y^2) plus the complement of the expected proportion $(1-p)$ times the square of the average recorded amount ($\bar{Y}^2$). The dot over the equal sign signifies the approximate nature of the equality.

The truth of this proposition follows from the following two relationships:

$$E_p\left(\sum_1^N D_j^2\right) = p \sum_1^N Y_j^2$$

and

$$E_p\left(\sum_1^N D_j\right)^2 = p(1-p)\sum_1^N Y_j^2 + p^2\left(\sum Y_j\right)^2.$$

Using these, the following relationship holds:

$$E_p\left[\sum_1^N D_j^2 - \frac{\left(\sum_1^N D_j\right)^2}{N}\right] = p^2\left[\sum_1^N Y_j^2 - \frac{\left(\sum_1^N Y_j\right)^2}{N}\right] + p(1-p)\sum_1^N Y_j^2\left(1 - \frac{1}{N}\right).$$

Dividing both sides by N, and neglecting the terms of order $1/N^2$, results in the proposition.

The population ratio (R) averaged over all possible realizations is simply $(1 - p)$. That is, $E_pR = (1 - p)$.

Proposition 2 $E_p(\sigma_R^2)$

$$\doteq p(1-p)\{\sigma_Y^2 + \bar{Y}^2\}\left[1 + \frac{1}{N}\left(\frac{\sigma_Y^2}{\bar{Y}^2} + \frac{4}{1+\left(\frac{\sigma_Y}{\bar{Y}}\right)^2} - \frac{G_1}{\frac{\bar{Y}}{\sigma_Y}\left[1+\left(\frac{\bar{Y}}{\sigma_Y}\right)^2\right]} - 5\right)\right].$$

This complex expression may be simplified for large values of N to read

$$E_p(\sigma_R^2) \doteq p(1-p)\{\sigma_Y^2 + \bar{Y}^2\}.$$

The proposition follows from the following relationships:

a. $$\sigma_R^2 = \frac{1}{N}\sum_{1}^{N}(X_j - RY_j)^2$$

$$= \frac{1}{N}[\sum X_j^2 - 2R\sum X_jY_j + R^2\sum Y_j^2].$$

b. $E_p \sum X_j^2 = (1-p)\sum Y_j^2.$

c. $$E_p\left(\frac{\sum X_j}{\sum Y_j}\right)(\sum X_jY_j) = p(1-p)\frac{\sum Y_j^3}{\sum Y_j} + (1-p)^2\sum Y_j^2.$$

d. $$E_pR^2 = p(1-p)\frac{\sum Y_j^2}{(\sum Y_j)^2} + (1-p)^2.$$

e. $$G_1 = \frac{1}{N}\frac{\sum(Y_j - \bar{Y})^3}{\sigma^3}$$ (Fisher's measure of skewness).

The population regression coefficient

$$B = \frac{\sum(X_j - \bar{X})(Y_j - \bar{Y})}{\sum(Y_j - \bar{Y})^2}$$

averaged over all possible realizations is also $(1-p)$. That is, $E_pB = (1-p)$.

Proposition 3 $E_p(\sigma_G^2) \leq p(1-p)(\sigma_Y^2 + \bar{Y}^2)$.

This proposition follows from previously cited relationships together with the fact that $(E_pB)^2 \leq E_pB^2$.

Each of these propositions can be extended to stratified designs. For the stratified difference estimate, proposition 1 may be interpreted as providing an approximation for the standard deviation within each stratum. In this case, the relationship would be as follows:

$$E_p(\sigma_D^2) \doteq p_i[\sigma_{Y_i}^2 + (1-p_i)\bar{Y}_i^2],$$

where the subscript i denotes the ith stratum. In practice, the auditor would use a common proportion (p_i) for all strata unless he had specific information concerning the proportion of differences within the several strata.

The proposition concerning the ratio estimator can be extended to the combined ratio estimator. This is most easily done for the situation where all the stratum proportions (p_i) are the same. The relationship is expressed as

$$\sigma_{RC_i}^2 \doteq p(1-p)\{\sigma_{Y_i}^2 + \bar{Y}_i^2\}.$$

This approximation neglects additional terms of order $1/N$, where N is the total population size. The relationship follows from the following three equalities:

$$(1)\quad E_p\left(\sum_j X_{ij}^2\right) = (1-p)\sum_j Y_{ij}^2.$$

$$(2)\quad E_p\left(R_c \sum_j X_{ij}Y_{ij}\right) = p(1-p)\frac{\sum_j Y_{ij}^3}{\sum_i \sum_j Y_{ij}} + (1-p)^2 \sum_j Y_{ij}^2.$$

$$(3)\quad E_p\left(R_c^2 \sum_j Y_{ij}^2\right) = p(1-p)\frac{\sum_i \sum_j Y_{ij}^2}{\left(\sum_i \sum_j Y_{ij}\right)^2} \sum_j Y_{ij}^2 + (1-p)^2 \sum_j Y_{ij}^2.$$

Likewise, the result for the regression estimator can be extended to the case of the combined regression estimate. This extension, however, has not been verified in detail because the algebraic manipulations appear to be so very complex. Nevertheless, the fact that combined regression is at least as efficient as the combined ratio indicates that any approximation for the combined ratio would also be an upper bound for the combined regression.

Appendix 5A

Adjustments When Achieved Precision Does Not Equal Planned Precision—Positive Approach

When planning, the auditor specifies his tolerable sampling risk (β), his tolerable risk of overauditing (α), and his measure of materiality (M). Using the positive approach, the planned precision (A) is given by the following formula:

$$A = \frac{Z_{\alpha/2}}{Z_{\alpha/2} + Z_\beta} M.$$

If the achieved precision (A') does not equal the planned precision, an adjustment may be made to preserve the planned sampling risk (β). The adjustment (A'') is chosen so that

$$M - A'' = A'\left(\frac{Z_\beta}{Z_{\alpha/2}}\right)$$

or

$$A'' = M - A'\left(\frac{Z_\beta}{Z_{\alpha/2}}\right).$$

When the achieved precision is based on a factor from Student's t-distribution, the formula used becomes

$$A'' = M - A'\left(\frac{t_\beta}{t_{\alpha/2}}\right),$$

where the selected t-factors correspond to the achieved degrees of freedom.

Using the adjusted precision A'', the auditor tests whether the estimated audited amount ($\hat{X}$) is within the decision interval ($Y - A''$ to $Y + A''$) or

$$Y - M + A'(t_\beta/t_{\alpha/2}) \leq \hat{X} \leq Y + M - A'(t_\beta/t_{\alpha/2}).$$

If the estimated audited amount ($\hat{X}$) does not satisfy the above relationship, the auditor concludes that there may be a material amount of error. Rewriting the inequality, it follows that the auditor decides there may be a material amount of error provided

$$|\hat{X} - Y| + A'(t_\beta/t_{\alpha/2}) > M.$$

This, of course, is the same inequality used in the negative approach because in that case $t_\beta = t_{\alpha/2}$.

To test the sensitivity of the auditor's decision with regard to the possible underestimation of the standard error, multiply $A'(t_\beta/t_{\alpha/2})$ by the factor $(1 + \omega)$ where ω represents the fraction of understatement. If

$$|\hat{X} - Y| + (1 + \omega)A'(t_\beta/t_{\alpha/2}) \leq M$$

for the selected value ω, then the auditor's conclusion that there is no material error stands even if the estimated standard error is 100ω percent too low. The computer program (VAREVA) tests the sensitivity with $\omega = .1, .2$, and $.3$ corresponding to a 10 percent, 20 percent, and 30 percent underestimation respectively.

Appendix 5B

Sensitivity of the Sampling Risk to the Estimated Standard Error

The following table shows the actual sampling risk (β) corresponding to a target sampling risk (β_0) and the ratio of the estimated standard error ($\hat{\sigma}(\hat{X})$) to the true standard error ($\sigma(\hat{X})$) when either the negative or the positive approach is used.

	Ratio of $\hat{\sigma}(\hat{X})$ to $\sigma(\hat{X})$				
Target β_0	.95	.90	.85	.80	.70
.025	.031	.039	.048	.058	.085
.05	.058	.068	.081	.093	.123
.10	.111	.125	.138	.154	.184
.15	.161	.174	.189	.203	.233
.20	.212	.224	.239	.251	.278
.25	.258	.271	.281	.295	.316
.30	.309	.316	.326	.337	.356
.40	.405	.409	.417	.421	.429

This table indicates the penalty for underestimating the standard error by various percentages. For example, when the estimated standard error is ten percent below the actual standard error (ratio = .90), the actual risk is .068 instead of the nominal .05.

Appendix 6A

Computing Sample Sizes for pps Samples

Let M represent a material amount of error, Q represent a small amount of error, and Y represent the total recorded amount for all sampling units with positive recorded amounts.

Select the tolerable sampling risk ($\beta = R_1$) for having the upper bound of monetary error less than M when, in fact, the monetary error equals M.

Select the tolerable risk of overauditing (α) for having the upper bound of monetary error greater than M when, in fact, the monetary error equals Q.

Let $P_0 = M/Y$, $P_1 = Q/Y$. Then the required sample size is the same as the sample size required for an attribute test for a threshold rate P_0, risk of unwarranted reliance $\beta = R_1$, and risk of overauditing α at P_1. The time sharing computer program ATSIZ2 can be used, or the following manual procedure.

The Poisson approximation may be used to find values of the sample size (n) and a critical number of monetary errors (k) by finding values (n, k) that satisfy the following two inequalities:

$$(1) \qquad e^{-nP_0} \sum_{0}^{k} \frac{(nP_0)^j}{j!} \leqslant \beta.$$

$$(2) \qquad 1 - e^{-nP_1} \sum_{0}^{k} \frac{(nP_1)^j}{j!} \leqslant \alpha.$$

The following tables facilitate this calculation.

		β									
		.01	.05	.07	.10	.15	.20	.25	.30	.40	.50
k	0	4.61	3.00	2.66	2.30	1.90	1.61	1.39	1.21	.92	.69
	1	6.64	4.74	4.34	3.89	3.38	3.00	2.69	2.44	2.02	1.68
	2	8.41	6.30	5.83	5.32	4.73	4.28	3.92	3.62	3.11	2.67
	3	10.0	7.75	7.25	6.68	6.02	5.52	5.11	4.76	3.18	3.67
	4	11.6	9.15	8.88	7.99	7.3	6.72	6.27	5.89	4.24	4.67
	5	13.1	10.5	10.21	9.27	8.5	7.91	7.42	7.01	6.29	5.67

		α				
		.01	.05	.10	.25	.50
k	0	.010	.051	.105	.288	.693
	1	.149	.355	.532	.961	1.68
	2	.436	.818	1.10	1.73	2.67
	3	.823	1.37	1.75	2.54	3.67
	4	1.28	1.97	2.43	3.37	4.67
	5	1.79	2.61	3.15	4.22	5.67

The procedure is as follows:

1. Set $k = 0$. For selected β, solve $nP_0 = x(k, \beta)$ where $x(k, \beta)$ is tabled value corresponding to row k, column β.
2. Set $k = 0$. For selected α, solve $n'P_1 = x(k, \alpha)$ where $x(k, \alpha)$ is tabled value corresponding to row k, column α.
3. If $n' \geqslant n$, stop and use the next integer larger than n as the sample size and k as the critical number. If $n' \leqslant n$, increase k by 1 and repeat steps 1, 2, and 3.

For example, suppose $M = \$150{,}000$, $Q = \$40{,}000$, and $Y = \$2{,}000{,}000$. $P_0 = .075(150{,}000/2{,}000{,}000)$ and $P_1 = .02(40{,}000/2{,}000{,}000)$. Set $\beta = .15$ and $\alpha = .05$.

1. Set $k = 0$

$$(.075)n = 1.90$$
$$n = 25.3$$
$$(.02)\, n' = .051$$
$$n' = 2.55$$

2. $n' < n$, so set $k = 1$ and repeat the steps

$$(.075)n = 3.38$$
$$n = 45.06$$
$$(.02)\, n' = .355$$
$$n' = 17.75$$

3. $n' < n$, so set $k = 2$ and repeat the steps

$$(.075)n = 4.73$$
$$n = 63.06$$
$$(.02)\, n' = .818$$
$$n' = 40.9$$

4. $n' < n$, so set $k = 3$ and repeat

$$(.075)n = 6.02$$
$$n = 80.26$$
$$(.02)\, n' = 1.37$$
$$n' = 68.5$$

5. $n' < n$, so set $k = 4$ and repeat

$$(.075)n = 7.3$$
$$n = 97.3$$
$$(.02)\, n' = 1.97$$
$$n' = 98.5$$

6. $n' > n$, so the required sample size is 98 and the critical number is 4.

Appendix 6B

pps Sampling for Attributes

Adapting pps sampling to attributes is easily accomplished. The auditor defines the attribute just as he would when using unrestricted random sampling. Typically, the attribute will be evidence of compliance deviation from some pertinent accounting control and the sampling unit will be a transaction. Whenever there is evidence of a compliance deviation, the dollar amount of the transaction is regarded as being in error. In symbols,

$$D_j = \begin{cases} Y_j & \text{if there is evidence of a compliance deviation;} \\ 0 & \text{otherwise.} \end{cases}$$

The ordinary pps estimator is $(1/n) \sum (D_j/Y_j)$ and with D_j defined in this way, the pps estimator is simply m/n, where m represents the number of sampling units containing evidence of a compliance deviation. This fraction (m/n) represents the estimated fraction of the total recorded amount (Y) associated with the compliance deviation.

Attribute tables or a computer program can be used to evaluate the results of a pps attribute sample. For a specified one-sided reliability (R_1), the auditor determines the achieved upper precision limit $(P_U(m))$ corresponding to the observed number of sampling units with compliance deviations (m). This represents the upper limit of the fraction of the total dollars that could have a compliance deviation at the stated reliability (R_1). Multiplying this fraction $(P_U(m))$ times the total recorded amount (Y) results in an upper limit on the dollar amount of transactions that could have compliance deviations.

Appendix 6C

Combining Two or More pps Samples

At times it may be necessary to select two or more independent pps samples from the same population. For example, sample one may be used to examine one source of monetary error and sample two may be used to examine a second source. In such circumstances, the auditor desires to combine the results from the two samples in order to ascertain whether there might be a material amount of monetary error from either cause. How should the samples be planned?

If possible, each sample should be planned using the same material amount (M), the same tolerable sampling risk (β) and the same tolerable risk of overauditing (α). This will, of course, result in the same sample size for each sample. Evaluation of the samples considered together is accomplished by pooling the observed monetary errors and regarding all the errors as having come from a single sample of size equal to the common sample size.

When it is not possible to use the same sample size, the auditor may still pool all the observed errors and regard them as having come from a sample of size equal to the smallest sample size of the separate samples. This will lead to a conservative conclusion.

Appendix 7

Separate Ratio Estimator and Separate Regression Estimator

Separate Ratio Estimator

The separate ratio estimator is formed by computing the ratio of the total sample audited amount within each stratum (x_i) to the total sample recorded amount within each stratum (y_i), extending this ratio (x_i/y_i) times the stratum total recorded amount (Y_i), and adding the results for all strata. This may be symbolically represented as

$$\hat{X}_{RS} = \sum \frac{x_i}{y_i} Y_i$$

$$= \sum \hat{R}_{Si} Y_i, \quad \text{where } \hat{R}_{Si} = \frac{x_i}{y_i}.$$

Alternatively, the total stratum difference (d_i) may be used to estimate the total population difference by means of the following formula

$$\hat{D}_{RS} = \sum \frac{d_i}{y_i} Y_i,$$

and, of course, the estimated total audited amount can be expressed as

$$\hat{X}_{RS} = Y + \hat{D}_{RS}.$$

The estimated standard error of the separate ratio estimator may be computed by using the following formula

$$\hat{\sigma}(\hat{X}_{RS}) = \sqrt{\sum N_i(N_i - n_i)\frac{S_{RSi}^2}{n_i}},$$

$$\text{where } S_{RSi} = \sqrt{\frac{\sum_j x_{ij}^2 + \hat{R}_{Si}^2 \sum_j y_{ij}^2 - 2\hat{R}_{Si} \sum_j x_{ij} y_{ij}}{n_i - 1}}.$$

The quantity S_{RSi} can also be computed using the observed sample differences ($d_{ij} = x_{ij} - y_{ij}$). In terms of the differences the formula is

$$S_{RSi} = \sqrt{\frac{\sum_j d_{ij}^2 + (\hat{R}_{Si} - 1)^2 \sum_j y_{ij}^2 - 2(\hat{R}_{Si} - 1) \sum_j d_{ij} y_{ij}}{n_i - 1}}.$$

When all, or nearly all, of the differences are expected to be overstatements, the sample may be planned by using the following approximation

$$\sigma^2_{RSi} \doteq p_i(1-p_i)[\sigma_{Yi}{}^2 + \bar{Y}_i{}^2]$$

where p_i represents the proportion of sampling units within the ith stratum that contain differences, σ_{Yi} is the standard deviation of the recorded amounts in the ith stratum, and $\bar{Y}_i$ is the average recorded amount in the ith stratum.

Separate Regression Estimator

The separate regression estimator is formed by computing a regression coefficient (b_i) for each stratum and using the following formula:

$$\hat{X}_{GS} = \sum N_i\bar{x}_i + \sum b_i(Y_i - N_i\bar{y}_i).$$

The stratum regression coefficient (b_i) is computed from the following formula:

$$b_i = \frac{\sum (x_{ij} - \bar{x}_i)(y_{ij} - \bar{y}_i)}{\sum (y_{ij} - \bar{y}_i)^2}.$$

This may also be expressed in terms of the observed differences as

$$b_i = 1 + \frac{\sum d_{ij}y_{ij} - n_i\bar{d}_i\bar{y}_i}{\sum (y_{ij} - \bar{y}_i)^2}.$$

The estimated standard error of the separate regression estimator may be computed using the following formula:

$$\hat{\sigma}(\hat{X}_{GS}) = \sqrt{\sum N_i(N_i - n_i)\frac{S^2_{GSi}}{n_i}},$$

where $$S_{GSi} = \sqrt{\frac{\sum x_{ij}{}^2 - n_i\bar{x}_i{}^2 - 2b_i(\sum x_{ij}y_{ij} - n_i\bar{x}_i\bar{y}_i) + b_i{}^2(\sum y_{ij}{}^2 - n_i\bar{y}_i{}^2)}{n_i - 1}}.$$

The quantity S_{GSi} can also be computed using the observed sample differences ($d_{ij} = x_{ij} - y_{ij}$). In terms of the differences the formula is,

$$S_{GSi} = \sqrt{\frac{\sum d_{ij}{}^2 + (1-b_i)^2 \sum y_{ij}{}^2 + 2(1-b_i)\sum d_{ij}y_{ij} - n_i[\bar{d}_i + (1-b_i)\bar{y}_i]}{n_i - 1}}.$$

References

[1] Aiken, H., et al. *Tables of the Cumulative Binomial Probability Distribution.* Cambridge, Massachusetts: Harvard University Press, 1955.

[2] American Institute of Certified Public Accountants. *An Auditor's Approach to Statistical Sampling.* New York: AICPA.
   Vol. 1 "An Introduction to Statistical Concepts and Estimation of Dollar Values" (1973)
   Vol. 2 "Sampling for Attributes: Estimation and Decision" (1974)
   Vol. 3 "Stratified Random Sampling" (1973)
   Vol. 5 "Ratio and Difference Estimation" (1972)
   Vol. 6 "Field Manual for Statistical Sampling" (1974)

[3] Anderson, R., and Teitlebaum, A. "Dollar-Unit Sampling: A Solution to the Audit Sampling Dilemma," *CA Magazine,* April 1973.

[4] Arkin, H. *Handbook of Sampling for Auditing and Accounting.* New York: McGraw-Hill Book Company, 1974.

[5] Cochran, W. *Sampling Techniques.* 2d ed. New York: John Wiley & Sons, 1963.

[6] Cyert, R., and Davidson, H. *Statistical Sampling for Accounting Information.* Englewood Cliffs, N.J.: Prentice-Hall, 1962.

[7] Deming, W. *Sample Design in Business Research.* New York: John Wiley & Sons, 1960.

[8] Dixon, W., and Massey, F., Jr. *Introduction to Statistical Analysis.* 3d ed. New York: McGraw-Hill Book Company, 1969.

[9] Elliott, R. "Basic Concepts of Statistics and Hypothesis Testing for Auditing." Chapter 9 in *Handbook of Modern Accounting.* S. Davidson and R. Weil, editors. New York: McGraw-Hill Book Company, 1977.

[10] Elliott, R., and Rogers, J. "Relating Statistical Sampling to Audit Objectives." *Journal of Accountancy,* July 1972.

[11] Fienberg, S.; Neter, J.; and Leitch, R. "Estimating the Total Overstatement Error in Accounting Populations." *Journal of the American Statistical Association,* June 1977.

[12] Hanson, M.; Hurwitz, W.; and Madow, W. *Sample Survey Methods and Theory.* New York: John Wiley & Sons, 1953.

[13] Heerden, A. van, "Statistical Sampling as a Means of Auditing." *Maandblad voor Accountancy en Bedrijfshuish oulkunde,* 1961.

[14] Kaplan, R. "Statistical Sampling in Auditing With Auxiliary Information Estimates." *Journal of Accounting Research,* Autumn 1973.

[15] Kaplan, R. "Sample Size Computations for Dollar-Unit Sampling." *Studies on Statistical Methodology in Auditing: Journal of Accounting Research* 13 (Supplement 1975): 126–33.

[16] Mautz, R. *Fundamentals of Auditing.* New York: John Wiley & Sons, 1964.

[17] Molina, E. *Poisson's Exponential Binomial Limit.* New York: D. Van Nostrand Company, Inc., 1942.

[18] Neter, J. and Loebbecke, J. *Behavior of Major Statistical Estimates in Sampling Accounting Populations.* New York: American Institute of Certified Public Accountants, 1975.

[19] Raj, D. *Sampling Theory.* New York: McGraw-Hill Book Company, 1968.

[20] Teitlebaum, A. "Dollar-Unit Sampling in Auditing," Faculty of Management Working Paper, McGill University, 1973.

[21] Vance, L. and Neter, J. *Statistical Sampling for Auditors and Accountants.* New York: John Wiley & Sons, 1956.

[22] Wallis, W. and Roberts, H. *Statistics: A New Approach.* Glencoe, Ill.: Free Press, 1959.

[23] Wilburn, A. "Stop-or-Go Sampling and How It Works," *Internal Auditor,* May 1968.

Selected Bibliography

The following is a selected bibliography of articles and books prepared by the Statistical Sampling Subcommittee. The articles and books were selected for the most part from those published after 1966 through the first quarter of 1976. Three criteria were used in the selection process: the article is available in a source commonly read by the accounting profession, the article treats some aspect of statistical sampling in auditing, and the article was judged to be of potential usefulness to an auditor in obtaining background information or in solving a specific problem.

Articles

Akresh, Abraham D. "Use of the Ratio Estimate in Statistical Sampling—A Case Study." *New York Certified Public Accountant,* March 1971, pp. 221–24.

Auditing case study involving use of the ratio estimate. Stimulates interest to study this technique.

Anderson, Rod, and Teitlebaum, A. D. "Dollar-Unit Sampling: A Solution to the Audit Sampling Dilemma." *CA Magazine* (formerly *Canadian Chartered Accountant*), April 1973, pp. 30–38.

Discusses dollar-unit sampling and presents the arguments in favor of widespread use of the technique. Avoids technical details.

Arkin, Herbert. "Statistical Sampling and Internal Control." *CPA Journal,* January 1976, pp. 15–18.

The author states a case for use of two-sided confidence limits in attribute tests of internal control.

Bedingfield, James P. "The Current State of Statistical Sampling and Auditing." *Journal of Accountancy,* December 1975, pp. 48–55.

Boatsman, James R., and Crooch, G. Michael. "An Example of Controlling the Risk of a Type II Error for Substantive Tests in Auditing." *Accounting Review,* July 1975, pp. 610–15.

Burstein, Herman. "The 'Ratio Estimate': A Useful Sampling Technique." *CPA Journal* (formerly *New York Certified Public Accountant*), November 1967, pp. 844–50.

A general discussion of the usefulness of ratio estimation through presentation of seven brief case studies where the technique was used.

Carmichael, D. R. "Tests of Transactions—Statistical and Otherwise." *Journal of Accountancy,* February 1968, pp. 36–40.

Presents a comprehensive discussion of the nature of audit sampling objectives and the choice among sampling techniques to best achieve the objectives.

Copeland, Ronald M., and Englebrecht, Ted D. "Statistical Sampling: An Uncertain Defense Against Legal Liability." *CPA Journal,* November 1975, pp. 23–27.

The authors believe statistical sampling may be a defense in the "legal liability arena," but not without some perplexities, difficulties, and vulnerability.

Davidson, H. J.; Neter, J.; and Petran, A. S. "Estimating the Liability for Unredeemed Stamps." *Journal of Accounting Research,* Autumn 1967, pp. 186–207.

Presents a useful technique to estimate the liability for unredeemed trading stamps although the technique might be useful in other areas such as transportation companies.

Deakin, Edward B., and Granof, Michael H. "Directing Audit Effort Using Regression Analysis." *The CPA Journal,* February 1976, pp. 29–33.

Expresses the view that the technique of multiple regression enables the auditor to substitute statistical judgment for intuition in selecting areas for intensive investigation. Authors believe the technique, if used with caution, can be used in conjunction with other audit procedures to reduce cost and increase effectiveness.

Deming, W. Edwards, and Grice, T. Nelson, Jr. "An Efficient Procedure for Audit of Accounts Receivable." *Management Accounting,* March 1970, pp. 17–21.

A study in the practical application of statistical theory to the audit of a trucking company's freight bills receivable.

Doney, Lloyd D. "What Statistics Can Offer Certified Public Accountants." *Journal of Accountancy,* April 1968, pp. 86–88.

Outline of activities of the certified public accountant and partial inventory of related statistical techniques applicable to each activity.

Elliott, Robert K., and Rogers, John R. "Relating Statistical Sampling to Audit Objectives." *Journal of Accountancy,* July 1972, pp. 46–55.

Presents a sampling plan that specifically controls both types of risk accepted by auditors when they make a decision based on a sample. The implications of not controlling both types of errors, when sampling, are illustrated.

Goodfellow, James L.; Loebbecke, James K.; and Neter, John. "Some Perspectives on CAV Sampling Plans." *CA Magazine,* October and November 1974, pp. 46–53.

Hall, William D. "Inventory Determinations by Means of Statistical Sampling Where Clients Have Perpetual Records." *Journal of Accountancy,* March 1967, pp. 65–66, 68–71.

Basic concepts in determining inventories by means of statistical sampling.

Hubbard, T. D., and Bullington, J. B. "Positive and Negative Confirmation Requests—A Test." *Journal of Accountancy,* March 1972, pp. 48–56.

The article presents a case study of a test of effectiveness of positive versus negative confirmation requests of receivable balances. Two samples of 102 items were selected. One-third of each were overstated, one-third were understated, and one-third were left unchanged. One group was sent negative requests and the second positive requests for confirmation. The results do not show that positive requests were more effective than negative. They do cast a shadow over the overall effectiveness of the audit procedures of confirmation.

Ijiri, Yuji, and Kaplan, Robert S. "The Four Objectives of Sampling in Auditing: Representative, Corrective, Protective, and Preventive." *Management Accounting,* December 1970, pp. 42–44.

Considerations in the design of sampling plans, both statistical and nonstatistical.

———. "A Model for Integrating Sampling Objectives." *Journal of Accounting Research,* Spring 1971, pp. 73–87.

Gives the objectives for audit application of statistical sampling and the methods of achieving the objectives.

Ilderton, Robert S. "Statistical Sampling Facilitated by DCAA Software Programs." *Defense Management Journal,* October 1975, pp. 34–38.

Explains a number of computer programs in use by the Defense Contract Audit Agency (and generally available to the public) for selecting and evaluating random samples. Programs include random number generation, ratio and difference estimation, dollar-unit sampling, and Bayesian methods.

Kaplan, Robert S. "Statistical Sampling in Auditing with Auxiliary Information Estimators." Report 7314, Graduate School of Business, University of Chicago. *Journal of Accounting Research,* Autumn 1973, pp. 238–58.

Discusses problems in variables sampling because of general low error rate in accounting population.

The article is an in-depth, sophisticated discussion of the advantages and usefulness of various estimators for use in variables estimation techniques.

———. "A Stochastic Model for Auditing." *Journal of Accounting Research,* Autumn 1973, pp. 38–46.

Kaufman, Stuart F. "Sampling for Zero Defectives." *Journal of Accountancy,* October 1968, pp. 66, 68–69.

Discusses availability of a table for determining sample size when zero errors are expected in an accounting population. The table is based on hypergeometric distributions.

Kern, C. L.; Neyhart, C. A., Jr.; and Hock, C. A. "Statistical Risk and Sample Reliability," *CPA Journal,* January 1973, pp. 55–60.

Kinney, William R. Jr. "A Decision-Theory Approach to the Sampling Problem in Auditing." *Journal of Accounting Research,* Spring 1975, pp. 117–32.

An argument for the use of decision theory over other dollar-value sampling approaches. The approach is compared with hypothesis testing and the traditional confidence interval approach.

Kraft, William H. Jr., "Statistical Sampling for Auditors: A New Look." *Journal of Accountancy,* August 1968, pp. 49–56.

Advocates use of Bayes formula to combine subjective assurance with sample results. Illustrates the calculations required to evaluate the auditors' combined assurance that the error rate in a population does not exceed a critical error rate.

Loebbecke, James K., and Neter, John. "Statistical Sampling in Confirming Receivables." *Journal of Accountancy,* June 1973, pp. 44–50.

Presents an approach to evaluating statistical samples using both positive and negative confirmation requests, including the role of alternative procedures.

Meikle, Giles R. "Statistical Sampling in an Audit Context." *Audit Technique Study,* Canadian Institute of Chartered Accountants, 1972, p. 43.

The study explains the basic sampling concepts (reliability levels and precision) and explains how these concepts fit into auditing theory. It describes the various sampling techniques that are available, such as attribute sampling and variables estimation sampling, and comments on the feasibility of their application in auditing. The study also warns of the most common problems encountered in the application of statistical sampling to auditing in areas such as sample design, selection of items for examination, error recognition, and sample evaluation. The pitfalls to avoid and how to avoid them are discussed.

Nigra, Alphonse L. "Statistical Sampling With EDP Equipment." *Internal Auditor,* March-April 1969, pp. 18–34.

Restall, Lawrence J., and Czajkowski, Peter J. "Computation of LIFO Index: A Statistical Sampling Approach." *Management Accounting,* September 1969, pp. 43–48.

Discusses the application of a statistical approach to arrive at an index for use in pricing inventories carried on a LIFO method.

Roberts, Donald M. "A Statistical Interpretation of SAP no. 54." *Journal of Accountancy,* March 1974, pp. 47–53.

A discussion of the implications of Appendix B—"Precision and Reliability for Statistical Sampling in Auditing" to "The Auditors Study and Evaluation of Internal Control" now included in *Codification of Statements on Auditing Standards* (American Institute of Certified Public Accountants, 1977).

———. "Sample Size Determination for Attributes." *Journal of Accountancy,* June 1975, pp. 46–48.

An answer to an inquiry concerning determination of sample size for an attribute sample using the table in an AICPA individual study program, "Sampling for Attributes: Estimation and Discovery."

Sawyer, Lawrence B. "The Lawyer, the Statistician, and the Internal Auditor." *Internal Auditor,* Summer 1967, pp. 9–18.

Presents an example of possible legal consequences of employing statistical sampling to determine the boundary of a variable. Stresses the need to plan, define, and document a sample.

———. "Simple Sampling: How to Stop Worrying and Learn to Love Statistical Tables." *Internal Auditor,* July-August 1968, pp. 9–26.

Well-illustrated discussion of basic concepts of statistical sampling that minimizes jargon and sets forth ten important principles for the auditor to learn.

Scott, William R. "Statistical Sampling in Auditing." *CA Magazine,* July 1975, pp. 13–17.

Smith, Kenneth A. "The Relationship of Internal Control Evaluation and Audit Sample Size." *Accounting Review,* April 1972, pp. 260–69.

Compares arbitrary variation of confidence levels and the Bayesian approach and concludes Bayesian is preferable.

Sorensen, James E. "Bayesian Analysis in Auditing." *Accounting Review,* July 1969, pp. 555–61.

An illustration and endorsement of possible applications of Bayesian analyses in auditing.

Taylor, Robert G. "Error Analysis in Audit Tests." *Journal of Accountancy,* May 1974, pp. 78, 80–82.

Discusses various types of errors that may be found while sampling and their disposition during the work.

Tracy, John A. "Bayesian Statistical Confidence Intervals for Auditors." *Journal of Accountancy,* July 1969, pp. 41–71.

Discusses the applications of Bayesian analysis to modify classic confidence intervals where attribute estimation is used. Illustrative data is included.

Tummens, Marvin, and Watson, Hugh J. "Advantages of Regression Analysis Over Ratio Analysis." *CPA Journal,* May 1975, pp. 35–38.

Uyeda, Susumu. "Statistical Sampling Procedures in Examination of Vouchers." *GAO Review,* Winter 1968.

Article discusses the development and implementation of a statistical sampling procedure to determine overpayment error in vouchers.

Warren, Carl S. "Interpreting and Evaluating Attribute Sampling." *Internal Auditor,* July-August 1975, pp. 45, 48–56.

Wilburn, Arthur J. "Stop-or-Go Sampling and How It Works." *Internal Auditor,* May-June 1968, pp. 10–20.

Describes the use of stop-or-go sampling including the use of tables for sample size determination and evaluation.

———. "Observations of a Statistician in an Auditing Environment." *Internal Auditor,* January-February 1974, pp. 56–62.

Discussion of some of the problems of applying statistical sampling to auditing due to some of the unique objectives of auditing and the "language barrier" which may result when statisticians and auditors do not completely understand each others work.

Yu, S., and Neter, John. "A Stochastic Model of the Internal Control System." *Journal of Accounting Research,* Autumn 1973, pp. 273–95.

Books

American Institute of Certified Public Accountants. *An Auditor's Approach to Statistical Sampling.* 6 vols. New York: AICPA, 1972–74.

Programmed learning texts designed for individual study. Include *An Introduction to Statistical Concepts and Estimation of Dollar Values* (vol. 1); *Sampling for Attributes* (vols. 2 and 4); *Stratified* (vol. 5); and *Field Manual for Statistical Sampling* (vol. 6). Useful for basic training programs for statistical sampling in auditing.

———. *Codification of Statements on Auditing Standards.* "Appendix A: Relationship of Statistical Sampling to Generally Accepted Auditing Standards" and "Appendix B: Precision and Reliability for Statistical Sampling in Auditing." New York: AICPA, 1977.

Arens, Alvin A., and Loebbecke, James K. *Auditing: An Approach.* Englewood Cliffs, N.J.: Prentice-Hall, 1976.

A comprehensive conceptual approach to auditing which is both pragmatic and innovative. Procedures and concepts are presented clearly, avoiding "how-to" prescriptions. Early treatment of statistical sampling, clear explanations of complex terminology and recent AICPA pronouncements are features of this text.

Arkin, Herbert. *Handbook of Sampling for Auditing and Accounting.* New York: McGraw-Hill Book Company, 1974.

One of the standard reference texts for the auditor or accountant who wishes to use statistics in his work. The book contains numerous tables and a lucid explanation of statistical formulas. A large number of statistical sampling plans and methods are also displayed.

Bernstein, Leonard A. *Statistics for the Executive.* New York: Hawthorn Books, 1970.

An introductory exposition of how statistical techniques may be useful to the executive decision-making process. The objective is to give the reader an appreciation of what statistics can and cannot do.

Brown, R. G., and Vance, L. L. *Sampling Tables for Estimating Error Rates of Proportions.* Berkeley, Calif.: University of California, 1961.

Burstein, Herman. *Attribute Sampling.* New York: McGraw-Hill Book Co., 1971.

———. *Sample-Size Tables for Quality Control and Auditing.* Westport, Conn.: Redgrave Information Resources Corporation, 1973.

Canadian Institute of Chartered Accountants. *Internal Control and Procedural Audit Tests.* Toronto: Canadian Institute of Chartered Accountants, 1968.

Cochran, William G. *Sampling Techniques.* 2d ed. New York: John Wiley & Sons, 1963.

Mathematical formula and proofs underlying most commonly used statistical sampling techniques.

Coen, Charles D. *Statistical Sampling for Bank Auditors.* Park Ridge, Ill.: Bank Administration Institute, 1973.

Intended for bank auditors, this booklet includes a review of basic concepts and various sampling methods. It includes an illustration of means of implementing sampling in a bank.

Cyert, Richard M., and Davidson, H. Justin. *Statistical Sampling for Accounting Information.* Englewood Cliffs, N.J.: Prentice-Hall, 1962.

General reference and learning text for statistical sampling methods commonly used in accounting and auditing. Brief and not overly technical. Problems and solutions are included.

Defense Contract Audit Agency. *General Purpose Time-Shared Computer Programs.* Memphis, Tenn.: Department of Defense, Defense Contract Audit Agency, 1976.

Listings, descriptions, and sample runs of time shared computer programs which have application to activities at a number of DCAA locations. Many programs may have application to activities of internal and external auditors.

Defliese, Philip L.; Johnson, Kenneth P.; and Macleod, Roderick K. *Montgomery's Auditing.* 9th ed. New York: Ronald Press, 1975, pp. 811–29.

As stated on page 811, "An auditor's use of statistical sampling in performing audit tests may be appropriate in certain circumstances . . . The purpose . . . is to provide additional guidance concerning the suitability and application of statistical sampling techniques. It is not intended as a full exposition of the theory and application of mathematical statistics. . . ." A well-written section concerning statistics in auditing by proponents of limited use of statistical techniques in auditing.

Dixon, Wilfred J., and Massey, Frank J., Jr. *Introduction to Statistical Analysis.* 3d ed. New York: McGraw-Hill Book Co., 1969.

A standard textbook introduction to the field of statistics. Accessible to those who have not had calculus. The book includes many useful tables and charts.

Hadley, G. *Probability and Statistical Decision Theory.* San Francisco: Holden-Day, 1967.

This text is a challenging introduction to applied decision theory primarily using business examples. The necessary elements of probability theory and utility theory are included. The last five of eleven chapters require some skill with calculus.

Institute of Internal Auditors. *Sampling Manual for Auditors* and *Supplement to Sampling Manual for Auditors.* Orlando, Fla.: Institute of Internal Auditors, 1967 and 1971.

Working handbooks on the application of basic sampling plans and techniques for auditors.

McRae, Thomas W. *Statistical Sampling for Audit and Control.* London: John Wiley & Sons, 1974.

Neter, John; Wasserman, William; and Whitmore, G. A. *Fundamental Statistics for Business and Economics.* 4th ed., Boston: Allyn and Bacon, 1973.

———, and Loebbecke, James K. *Behavior of Major Statistical Estimators in Sampling Accounting Populations.* New York: AICPA, 1975.

An empirical investigation of a variety of important complex problems in the use of major statistical estimators in accounting populations.

Newman, Maurice S. *Financial Accounting Estimates through Statistical Sampling by Computer.* New York: John Wiley & Sons, 1976.

An exposition of some standard statistical estimation techniques together with many suggestions for their use in producing estimates for financial and accounting purposes. A set of computer programs is discussed and illustrated.

Pearson, E. S., and Hartley, H. O. *Brometrika Tables for Statisticians.* vol. 1. London: Cambridge, 1966.

Raj, Des. *Sampling Theory.* New York: McGraw-Hill Book Co., 1968.

———. *The Design of Sample Surveys.* New York: McGraw-Hill Book Co., 1972.

Texts on classical sampling theory and methods. Quite thorough and includes a considerable amount of derivations requiring calculus. A well-equipped reader will find these texts to be good references.

Vanassee, Robert W. *Statistical Sampling for Auditing and Accounting Decisions: A Simulation.* New York: McGraw-Hill Book Co., 1968.

Intended for training exercises. Includes a brief summary of attribute, variable (mean-per-unit), and discovery sampling. Exercises on each technique using simulated populations are the principal focus of this text.

Wannacott and Wannacott. *Introductory Statistics.* New York: John Wiley & Sons, 1969.

Williams, Thomas H., and Griffin, Charles H. *Management Information, A Quantitative Accent.* Homewood, Ill.: Richard D. Irwin, 1967.

Willingham, John, J., and Carmichael, D. R. *Auditing Concepts and Methods.* 2d ed. New York: McGraw-Hill Book Co., 1975.

Develops a uniformly conceptual approach to the teaching of auditing.

Yamane, Taro. *Elementary Sampling Theory.* Englewood Cliffs, N.J.: Prentice-Hall, 1967.

Introduction to sampling theory. Written for undergraduate students with a mathematics or statistics background, it covers the basic concepts and most types of sampling usually encountered.

Index